NEW PERSPECTIVES IN INDIAN SCIENCE AND CIVILIZATION

This book examines key aspects of the history, philosophy, and culture of science in India, especially as they may be comprehended in the larger idea of an Indian civilization. The authors, drawn from a range of disciplines, discuss a wide array of issues – scientism and religious dogma, dialectics of faith and knowledge, science under colonial conditions, science and study of grammar, western science and classical systems of logic, metaphysics and methodology, and science and spirituality in the *Mahabharata*. This collection of essays aims to evolve a framework in which science, culture, and society in India may be studied fruitfully across disciplines and historical periods.

With its diverse themes and original approaches, the book will be of interest to scholars and researchers in the fields of the history and philosophy of science, science and religion, cultural studies and colonial studies, and philosophy and history, as well as India studies and South Asian studies.

Makarand R. Paranjape is Director, Indian Institute of Advanced Study, Shimla, India. He has been Professor of English, Jawaharlal Nehru University, New Delhi, India, for close to twenty years. Earlier, he taught at the Indian Institute of Technology-Delhi and the University of Hyderabad. His overseas assignments include the Eric Auerbach Visiting Chair in Global Literary Studies at the University of Tubingen, Germany and the ICCR Chair in Indian Studies at the National University of Singapore. He was educated at St. Stephen's College, University of Delhi (BA Hons.) and received his master's and doctoral degrees at the University of Illinois at Urbana-Champaign, USA. He is the author/editor of over 45 books which include works of criticism, poetry and fiction and has published over 175 academic papers. His latest works include *Debating the 'Post' Condition in India: Critical Vernaculars, Unauthorized Modernities, Post-Colonial Contentions* (2017), *Cultural Politics in Modern India* (2016), *The Death and Afterlife of Mahatma Gandhi* (2015), *Making India: Colonialism, National Culture, and Indian English Literature* (2015), and *Acts of Faith: Journeys to Sacred India* (2012).

NEW PERSPECTIVES IN INDIAN SCIENCE AND CIVILIZATION

Edited by Makarand R. Paranjape

Routledge
Taylor & Francis Group

LONDON AND NEW YORK

First published 2020
by Routledge
2 Park Square, Milton Park, Abingdon, Oxon OX14 4RN

and by Routledge
605 Third Avenue, New York, NY 10017

First issued in paperback 2021

Routledge is an imprint of the Taylor & Francis Group, an informa business

British Library Cataloguing-in-Publication Data
A catalogue record for this book is available from the British Library

Library of Congress Cataloging-in-Publication Data
Names: Paranjape, Makarand R., 1960– editor.
Title: New perspectives in Indian science and civilisation / edited by
Makarand R. Paranjape.
Description: Abingdon, Oxon ; New York, NY : Routledge, 2019. | Includes
bibliographical references.
Identifiers: LCCN 2018057730 | ISBN 9781138342859 (hardback : alk. paper) |
ISBN 9780429244148 (e-book)
Subjects: LCSH: Science—India. | Science—India—History. | Religion and science—India. |
Religion and science—India—History. | Science—Social aspects—India. |
Science and civilization.
Classification: LCC Q127.I4 N49 2019 | DDC 509.54—dc23
LC record available at https://lccn.loc.gov/2018057730

Typeset in Sabon
by Apex CoVantage LLC

ISBN 13: 978-0-367-78456-0 (pbk)
ISBN 13: 978-1-138-34285-9 (hbk)

To two awesome twosomes,
Medha and Sharad Deshpande
Bhagyada and Bhushan Patwardhan

CONTENTS

CONTENTS

CONTRIBUTORS

Sitansu Chakravarti holds a PhD in philosophy from Syracuse, New York, USA. Formerly Visiting Professor in Comparative Religion, Visva-Bharati, Santiniketan, and Visiting Professor in Philosophy, University of Rajasthan, he is the author of *Ethics in the Mahabharata: A Philosophical Inquiry for Today* (2006).

Madhumita Chattopadhyay is Professor of Philosophy at Jadavpur University, Kolkata, India. Her areas of specialization are Buddhist philosophy, logic, contemporary Indian philosophy, Indian ethics. She has received several awards and scholarships, including a BDK Fellowship from Japan, a Fulbright Visiting Fellowship, and a JSPS Fellowship from the Japanese Society for the Promotion of Science. She has published books such as *Walking along the paths of Buddhist Epistemology* (2008) and *Interpersonal Relations in Academy* (2007).

Probal Dasgupta was Professor of Linguistics, Indian Statistical Institute, Kolkata. He completed his PhD in linguistics in 1980 from New York University, USA. A former instructor in linguistics in many international universities, including New York, Melbourne, and Barlaston College, he has also been Professor of Linguistics, University of Hyderabad. His recent publications include *After Etymology (2000)* and *Inhabiting Human Languages: the Substantivist Visualization* (2012).

Professor Sharad Deshpande's teaching and research career spans over four decades in the core areas of philosophy, particularly in 20 th Century Indian Philosophy, Analytic Philosophy, Philosophy of History, and Philosophy of Action. He was Tagore Fellow at Indian Institute of Advanced Study, Shimla, in 2012–2014, and Professor of Philosophy at S.P. Pune University.His publications include *Philosophy in Colonial India* (2015), *Philosophy of G.R.Malkani* (1997), and *200 Years of Kant* (2004) besides a vast number of research articles in various journals and edited volumes.

Nirmalya Guha is Associate Professor, Indian Institute of Technology, Banaras Hindu University, Varanasi, India. He was previously Associate Professor at the Manipal Centre for Philosophy and Humanities, and Assistant Professor in the Department of Humanities and Social Sciences, IIT-Kanpur, India. He holds an MA in philosophy from the University of Madras (2003), an MA in applied linguistics from the Central University of Hyderabad (2000), and PhD from Lancaster University, U.K. The topic of his doctoral research was *Postulation: Arthāpatti in Indian Epistemology*. He has published many articles in various journals such as *Journal of Indian Council of Philosophical Research* and *Indian Philosophical Quarterly*.

Job Kozhamthadam was President of Jnana-Deepa Vidyapeeth, Pune, India, where he was also Professor of Philosophy of Science. He is a member of the International Society for Science and Religion (ISSR), Cambridge, UK, former member of the Indian National Commission for the History of Science, INSA, and the founder-director of IISR (Indian Institute of Science and Religion). His book *The Discovery of Kepler's Laws: The Interaction of Science, Philosophy and Religion* (1994) was selected by *Choice Magazine* as one of the outstanding academic books of the year 1994.

Sampadananda Mishra is Director of Sri Aurobindo Foundation for Indian Culture (SAFIC), Sri Aurobindo Society, Puducherry, India. He was initiated into Sanskrit by his grandfather, Pandit Paramananda Mishra. In 1993, after finishing his postgraduation in Sanskrit from Utkal University, Bhubaneswar, he completed his MPhil in Sanskrit at Pondicherry Central University. He has published books such as *Chandovallari* (1999) and *The Wonder That Is Sanskrit* (2002).

Sreekala M. Nair was Professor and Head, Department of Philosophy, Sree Sankaracharya University of Sanskrit, Kalady, Kerala. She acquired her PhD from the University of Madras on contemporary epistemology in 1994 and joined the Sankaracharya University in the same year. Her areas of specializations are contemporary epistemological issues, contemporary analytic metaphysics, epistemic logic, Indian epistemology, consciousness studies (Indian), cultural studies (Indian), Indian aesthetics, and so on. She has published around fifty research papers in different journals and compiled volumes.

Augustine Thomas Pamplany is the Founder-Director of the Institute of Science and Religion at Little Flower Seminary, Aluva, India. He is the Managing Editor and Publisher of *Omega – Indian Journal of Science and Religion*. He holds a PhD in philosophy of science from the International University, Colombo, for his thesis on *The Metaphysics of Quantum Physics*. Some of his publications include *Theological Mysteries in Scientific Perspective* (2005); *Cosmos, Bios, and Theos: Introduction to Philosophy of Science, Scientific Cosmology and Science-Religion*

Dialogue (2004); and *East-West Interface of Reality: A Scientific and Intuitive Inquiry into the Nature of Reality* (2003) co-authored with Dr. Job Kozhamthadam.

Bhushan Patwardhan has over thirty years of experience in research and development in evidence-based Ayurveda, ethnopharmacology, drug discovery and development, and integrative medicine. He is the Vice-Chairman of UGC, Fellow of the National Academy of Medical Sciences (India), and Chairman of Academic Planning and Development Committee of National Institute of Pharmaceutical Education & Research. He has been academic head of Manipal Education Group; Director, Institute of Ayurveda and Integrative Medicine, Bengaluru; and Vice Chancellor of Symbiosis University. He is editor-in-chief of *Journal of Ayurveda and Integrative Medicine* and the author of several books and academic papers. His latest major book-length publications include *Integrative Approaches for Health: Biomedical Research, Ayurveda and Yoga* (2015) and *Innovative Approaches in Drug Discovery: Ethnopharmacology, Systems Biology and Holistic Targeting* (2016).

V. V. Raman is Emeritus Professor of Physics and Humanities at the Rochester Institute of Technology, USA, and has an abiding interest in Hindu traditions and culture. He has contributed papers to international conferences and is a lecturer. His publications include *Truth and Tension in Science and Religion* (2009) and *Indic Visions: In an Age of Science* (2011).

K. Ramakrishna Rao, former Chairman, Indian Council for Philosophical Research, is a psychologist and educationist with vast experience in national and international arenas as a teacher, researcher and administrator. He was Chancellor, GITAM deemed University, Vishakhapatnam, India. The government of India awarded him the civilian honour of Padma Shri in 2011. He studied philosophy under the tutelage of Professors Saileswar Sen and Satchidananda Murthy at Andhra University and Richard McKeon at the University of Chicago, and received his PhD and DLit degrees. He has authored books such as *Consciousness Studies: Cross-cultural Perspectives* (2002) and *The Basic Experiments in Parapsychology* (1984).

Swami Sarvabhutananda was Secretary, Ramakrishna Mission Institute of Culture, Gol Park, Kolkata. He offered these opening remarks in the conference in which several of these chapters were first presented as papers.

Joy Sen is Professor, Department of Architecture & Regional Planning, Indian Institute of Technology, Kharagpur, and holds a PhD from there. Co-author of *A Systems Evaluation of Global History of Indian Architecture* (2016), he has published many scholarly articles, including "The

Price of Living" (2012) and "Epistemology and Ontology of Retracing the Paradigm of Evolution and Involution in Indic Studies" (2014).

Debabrata Sen Sharma was Professor in Sanskrit at Kurukshetra University and completed a PhD from Banaras Hindu University earlier. His areas of specialization are Indian philosophy and religion, and Kashmir Saivism. He has published more than sixty research articles in different research journals and six books on Kashmir Saivism and Tantra, including *Aspects of Tantra Yoga* (2007) and *Introduction to the Advaita Saiva Philosophy of Kashmir* (2009).

R. P. Singh is Professor at the Centre for Philosophy, School of Social Sciences, Jawaharlal Nehru University, New Delhi, India. His academic and research interests include modern Western philosophy, postmodern philosophy, and Indian philosophy. He has published over sixteen books and ninety-five research articles in academic journals.

Raman Srinivasan was educated at IIT-Madras (BTech) and the University of Pennsylvania, where he completed a PhD in interdisciplinary research on the history of science and consciousness in India. He worked for the Rockefeller Foundation before returning to India. He now serves in a senior position at Tata Consultancy Services.

FOREWORD

This collection of sixteen essays on *New Perspectives in Indian Science and Civilization* is, to say the least, uncommon. An uncommon pursuit is a deviation from the routine which most human activity – be it science, art, religion, or even the most of everyday activity – ends up as. Such deviations are intentional and not accidental. Though infrequent, they are not erratic. They are prompted by a thirst to know what is previously unknown. They are meant to cross the boundaries that characterize the routine. What these deviations result in are the turning points that change the course of the concerned human activity and of human life in general. The significance of such uncommon pursuits therefore lies in their emancipatory outcomes. Of such uncommon pursuits there cannot be any *a priori* theory. What is needed therefore is a phenomenology of uncommon pursuits – a phenomenology of deviations.

The essays in this volume seek to comprehend science, i.e., modern science, and *civilization* – the term very thoughtfully used by the editor of the volume – in an Indian perspective. This perspective is shaped by two considerations, namely, to know what science has been and to ask what science means, especially in the Indian context. Any attempt of this kind that asks these questions will be counted as an uncommon pursuit as it problematizes the dominant and the widespread view that science is its own authority and that it requires no independent validation. Despite its claim to universality and objectivity implying that science stands by itself, the fact remains, as the editor of this volume states, that science cannot exist in a vacuum. Despite being the most dominating force, modern science is but a part in the totality of human civilization. A critical investigation into the nature of this relationship between the part and the whole always opens up new possibilities of understanding both – the part and the whole. This volume is a significant step in exploring such possibilities.

It is undeniably true that, from its inception, the modern science to which almost all parts of the globe are exposed is a new element in human experience. It is a new and unprecedented form of knowledge that has come under philosophical scrutiny right from its beginning because it gives a new account of the universe as governed by a system of laws which are independent of God or Man. The ramifications of this account have been far-reaching: e.g., challenging the validity of its own ancestry, which comprised all the medieval cosmologies, knowledge systems, methods of reasoning, structures of causality, and canons of authority. This new account of the universe also problematized its metaphysical status, thereby introducing a separation of knowledge from value.

The essays in this volume dwell upon some of these ramifications by way of reflecting on the complex relationship between science and spirituality as well as science and culture in the context of Indian civilization. This is a distinctive feature of this volume, as it stresses that the reception of modern Western science in Indian cultural milieu was bound to raise issues of civilizational significance, of self-identity, and of cosmopolitan intent. Thus, one finds the contributors of this volume discussing a range of views and ideas in the framework of the distinctiveness of Indian culture and its indigenous evolution. Some of the issues discussed include scientism and religious dogmas, dialectics of faith and knowledge, pursuit of science under colonial conditions, science and study of grammar, Western science and Indian systems of logic, metaphysics and methodology, and classical texts like the *Mahabharata*. This diversity of issues coupled with originality of approach makes this volume not only interesting to read but demanding of our attention.

Sharad Deshpande

PROLOGUE

Indian Psyche and Science

The question has often been raised as to whether the Indian psyche and culture is conducive to the growth and development of science. We can answer that with an emphatic "yes."

In the Indian tradition, we come across two distinct terms – *rishi* and *muni*. The scriptures say that *risher darshanat* – a seer of super sensual truth – is a *rishi*, and *muner mananat* – a man who usually contemplates about things which we consider secular today – is a *muni*. The former pertains to *para vidya* – divine knowledge – and the latter to *apara vidya* – secular knowledge. Philosophers, researchers in fine arts and natural sciences, and sometimes even poets were considered *munis*. So, in its essence Indian tradition has always been reverential of any science, speculative or hard-core.

Swami Vivekananda said:

> in reality all this difference [between *para* and *aparavidya*] is only one of degree and not of kind. It is not that secular and spiritual knowledge are two opposite and contradictory things; but they are the same thing – the same infinite knowledge which is everywhere fully present from the lowest atom to the highest Brahman – they are the same knowledge in its different stages of gradual development.
>
> (434)

Swami Vivekananda also pointed out the characteristic feature of the Indian concept of science. Indians sought to explore the inner world and felt that the outer universe is the projection of the inner universe. They searched for the truth within and then saw its reflection without. This relationship between macrocosm and microcosm was reflected in all Indian thought. The Siddha and Hathayoga philosophy popularized the concept that the human body itself contains the entire universe in it and that the microcosm and macrocosm work on the same principles, only on different scales. It is on this basis that a school of science called *Rasayan Shastra*, similar to the alchemists of Europe, flourished.

The history of India is replete with scientific achievements. We find references to shipbuilding, making artificial limbs, concepts of geometric constructions, and even rules similar to Pythagorean triples and other scientific techniques in the *Shatapatha Brahmana* and the *Sulbasutras* well documented centuries before Christ. There are Vedic texts containing rudimentary algebra, geometry, trigonometry, primitive astronomy, and even veterinary medicine.

Contributions made by Shushruta and Charaka in medical science is well known, and it is believed that Shushruta even devised ways of surgery, especially cataract surgery. The Iron Pillar of Delhi and paintings of Ajanta are wondrous examples of the use of scientific techniques to promote art and culture.

It was in India that the zero was believed to have been discovered and the system of place-value numeration was developed, almost reaching to their fullest formulation by 500 CE. None outside of the Indian civilization had either devised such a complete system of numerical operation or evolved the zero-concept to the degree that it is used as the null-value in all facets of calculation.

We are all familiar with the names of Aryabhatta, Brahmagupta, and Varahamihira in ancient India. But we also do have Bodhayana, the Vedic scholar who worked on geometry; Panini, the grammarian who worked in the fields of phonetics, phonology, and morphology; and Kanada, the philosopher who proposed the atomic theory. A mention must be made of Pingala, the musical theorist, in whose work we find evidences of Pascal's triangle, binomial coefficients, binary numbers, and the like. Some developments were made in finding the infinite series for pi, an early version of the mean value theorem, series expansion for trigonometric functions, and other mathematical concepts.

In modern times, we see a revival of science in India. Pioneering new developments, Mahendranath Sircar, J. C. Bose, and P. C. Roy paved the way for the furtherance of science in the pre-independence era and were followed by the likes of Satyendra Nath Bose, Srinivasa Ramanujan, Meghnad Saha, C. V. Raman, Vikram Sarabhai, Homi Bhabha, Hargobind Khurana, and S. Chandrashekhar.

We might conclude this brief overview by suggesting that the development of scientific temper was actually achieved in post-independent India. Science was then no longer "Indian" or "Western," as it used to be; rather, it began to be referred to as simply "science," and this ushered in the Scientific Era.

Swami Sarvabhutananda

Works cited

Vivekananda, Swami. *The Complete Works of Swami Vivekananda*, Mayavati Memorial ed., Vol. IV. Kolkata: Advaita Ashrama, 2001.

ACKNOWLEDGEMENTS

The publication of this fourth and final volume in the series "Science and Spirituality in India" has been unusually long drawn. For a variety of personal and professional reasons, the journey to its completion has been painful and turbulent. I am glad at last to see the proverbial light at the end of what has been a very long and dark tunnel. Through this ordeal, I have contracted many debts of gratitude which I am happy to acknowledge. First of all, the contributors of this volume for their inordinate, almost superhuman, patience – thank you for sticking with the project, even if the publication of your chapter means considerably less to some of you now.

I am deeply grateful to those who supported the conference in which many of these chapters were originally presented. The Templeton Foundation–funded Special Awards Programme was led by two outstanding scholar-administrators, Professor Pranab Das and Mr. Tom Mackenzie. Their sustenance and encouragement have meant a lot to me through the bleak years. The Indian Council of Philosophical Research (ICPR), then led by Professor Ramakrishna Rao, who gave the keynote address, partnered with us. The Ramakrishna Mission Institute of Culture (RMIC) at Golpark, carrying forward Swami Vivekananda's legacy of exploring the relationship between science and religion, was the venue of the conference. The highly learned monks of the order helped and blessed us. RMIC even offered to publish the volume when we faced glitches. For this and much more, I wish personally to thank Swami Suparnanada-ji and Swami Yadavendra-ji. Jawaharlal Nehru University, where I've taught for the last twenty years, not only hosted my project but also offered logistical and administrative support. My then-research and project assistants, especially Sureshika, also deserve a special word of thanks for their unstinting help even when the going got tough. I am grateful to the various copyeditors, especially Surabhi, who went through the text, with them as also all those who helped in its publication. My continuing thanks to Shoma, Aakash, Rimina, and the great editorial team, both in acquisitions and production, at Routledge. This sixth collaboration is ample testimony not only of their competence but also to their continuing belief in me. Finally, my special thanks to the two most important ladies in my life: my mother, who has stood by me all these years, and Gayatri, who I now know as my first and last resort.

MRP
28 July 2019

1

INTRODUCTION

Makarand R. Paranjape

The purpose of this volume is to examine some aspects of the history, philosophy, and culture of science in India, all of which may be encompassed in the broader notion of civilization. Historians have shown that what we today call "science" was earlier known as "natural philosophy" in Europe. The term "philosophy" was then considered a generic description of all forms of human inquiry. Science, in other words, simply meant knowledge. The word "natural" was prefixed to the more general term philosophy to indicate that here the study was of nature and natural phenomena as opposed to other branches of inquiry which included grammar, rhetoric, literature, fine arts, theology, religion, metaphysics, theology, and so on. In India, too, there was no special word for science in pre-modern times. *Vidya, kala, jnana, vijnana, shiksha* – these and other such terms were used to denote what we today consider different branches of scientific knowledge.

The emergence of modern science occurred during and after the European Renaissance, which also witnessed a huge expansion of European power, both economic and political. In large measure, this expansion was in turn fuelled by new technologies. While many European nations expanded outward, colonizing the rest of the world, there was also fierce competition within Europe for survival and dominance. In such an environment, navigation, shipping, military, engineering, and technological superiority became crucial. The colonies, in turn, yielded gold, silver, silk, cotton, opium, spices, and a whole range of other natural resources. Maritime traffic facilitated the flow of knowledge and ideas from China, India, the Middle East, Africa, and the Americas to Europe and vice-versa. Finally, in the early eighteenth century, conditions were ripe for the Industrial Revolution in Europe. The Enlightenment had created a new thirst to know what was previously unknown: as Kant famously said, *sapere aude*, "Dare to know." But more than that, it also enthroned rationality as the highest faculty of the human being, thus paving the way for the retreat of Church dogma as the sole arbiter of truth. Instead of religious authority, empiricism and rationalism became the two dominant epistemological principles.

The aim of this broad sketch of the conditions that gave rise to modern science has been to show that it does not emerge in a vacuum but needs certain preconditions. A definite set of cultural conditions and circumstances give rise to modern science. Here, by culture is meant the sum total of all the devices and methods that a society has at its disposal to control, direct, and modify the material conditions of its existence. Once science begins to flourish, it gives rise to its own culture, both in the specific sense of the culture of laboratories or research institutions in which science is produced and, in the larger sense, a cultural ecosystem in which science begins to dominate society as a whole. The culture of science is thus an important aspect of study. It not only includes the study of the relationship between science and religion but also between science and philosophy and science and society. At the same time, the concept of culture needs to be revisited, anchored as it is increasingly in our time in the material bases of civilization. Today, culture relates itself more to the "human condition" and hence also to the "human predicament," understood in terms of the misery that man creates for himself externally and therefore must suffer the backlash for internally. Can science or scientific culture mitigate this predicament?

To come to the third crucial theme in the volume, by the culture of science, we also mean the logical, methodological, and epistemological assumptions of science as it was practised in India. This encompasses the clash or the uneasy coexistence between what we might call the "traditional" culture of India and what is now dominant and universal, that is, modern science. Here the crucial question is whether pre-modern science can be called science at all or simply an alternate knowledge system. But by science in India we also mean knowledge systems practiced in India before the advent of modernity, as well as science in colonial and post-colonial India. Similarly, by history we mean how science and scientific ideas came to be narrated in modern India, just as by culture we mean the larger cultural climate in the country over the centuries within which science developed.

This volume is the outcome of a major interdisciplinary conference held in Kolkata some ten years back. This conference, as well as three others prior to it, were organized under the auspices of an international project on "Indian Perspectives in Science and Spirituality" located at Jawaharlal Nehru University (JNU), New Delhi. All three of these conferences resulted in published volumes.

The first of these conferences, on "Science and Spirituality in Modern India," was held in February 2006 at the India International Centre and at JNU. With a special message of encouragement from the then president of India, Dr. A. P. J. Abdul Kalam, and inaugurated by His Holiness, the Dalai Lama, this conference brought together several luminaries from India and abroad from various disciplines. Speakers included stalwarts such as Dr. Karan Singh, Professor D. P. Chattopadhyay, Dr. Kireet Joshi, Swami Gokulananda, Shrivatsa Goswami, Professor V. V. Raman, Dr. Sudhir

Kakar, Professor P. L. Dhar, and so on. The conference was supported by Global Perspectives in Science and Spirituality, Indian Council of Cultural Relations, and several other organizations such as JNU, IIC, ICSSR, and ICPR. A special issue of *Life Positive* magazine was released ahead of the conference to highlight the importance of the topic to a wider audience. Two scholarly books, *Science and Spirituality in Modern India* (2007) and *Science, Spirituality, and the Modernization of India* (2008), were published as a result. The latter was selected as one of the volumes of the International Society of Science and Religion (ISSR) global library of the 150 most important books in the area and sent to over a hundred libraries worldwide.

The second conference was on "Information as Science and Spirituality." It explored the possibility of information being the link not only between various disciplines but also between science and spirituality. This event too invited a unique combination of artists, scientists, social scientists, philosophers, and literary critics. Held at the India Habitat Centre in March 2007, it was organized to coincide with the installation of a magnificent and gigantic bust of Shiva by leading artist Satish Gupta. The collaborators included the Visual Arts Gallery of the India Habitat Centre and the UGC Special Assistance Programme of the Centre for English Studies, JNU. A special issue of the quarterly *I* magazine of the India Habitat Centre was released to coincide with the symposium. This issue had an interview with the Tai Situpa, Pema Donyö Nyingche Wangpo, the teacher of the second-most revered person in Tibetan Buddhism, the Karmapa. It was inaugurated by the late Swami Veda Bharati, an eminent scholar and teacher of Vedanta, and the valedictory address was given by Dr. Karan Singh.

The third conference was on "Science and Spirituality in Healing." Held just outside Coimbatore, in Anaikatti, at the ashram of the late Swami Dayananda Saraswati in January 2008, the chief collaborator of the project on Indian Perspectives in Science and Spirituality was the Arya Vaidya Pharmacy (Coimbatore). The latter not only is a reputed manufacturer of Ayurvedic medicine but runs a hospital and research programme in the area. Most of the participants were health professionals, including allopathic doctors, Ayurvedic practitioners, and homeopaths, in addition to philosophers and social scientists. The outcome was the volume *Healing across Boundaries: Biomedicine and Alternative Therapeutics* published internationally by Routledge in 2015.

The present volume is based on a conference held in March 2009 at the Ramakrishna Mission Institute of Culture, Kolkata. Supported by the Indian Council of Philosophical Research (ICPR), this conference started with a keynote address by Professor Ramakrishna Rao, then Chair, ICPR, and had eminent scholars drawn from some of the best universities and institutions of India. The present collection, based on that conference, takes the debate on science and society to another area, that of culture and history, broadly included in the term civilization. It aims at opening up an interdisciplinary field of inquiry on fundamental issues that involve science and society.

The range of our concerns include the culture, history, and philosophy of science in India. The assumption is that science is either supported or resisted by the society and culture in which it functions. Of course, as suggested earlier, science has its own culture just as each culture has its own "science." Yet by culture of science is meant specifically four things: (1) the culture out of which science emerges, as in the European Renaissance or late colonial India; (2) the kind of culture that science itself engenders once it begins to take root; (3) the interaction between the two, that is between science, with its own specific norms and methodologies, and the culture of the country or society in which it exists; and (4) the institutionalization of science and its relationship to the state, which in turns shape the type of science that is encouraged.

Some of the topics and areas of concern include:

- Science as culture
- The culture of science in India, both institutional and personal
- Philosophy of science in India, past and present
- "Traditional" science in India: methodology and philosophy
- Traditional vs. modern science in India
- Science and society in India
- The aesthetics of science
- History and historiography of science in India

All told, the volume contains some outstanding essays by leading scholars drawn from different fields, including philosophy, religion, psychology, history, cultural studies, and of course, science and technology.

The volume opens with a prologue on "Indian Psyche and Science" by Swami Sarvabhutananda, then secretary of the Ramakrishna Mission Institute of Culture, the venue of the conference. He asks the important question whether the Indian psyche is conducive to the growth and development of science, answering it with an emphatic "yes." His main authority on this topic is not just the ancient scriptures of the Hindus, which clearly endorsed rational inquiry, but also, more recently, the work of Swami Vivekananda, an ardent proponent of modern science. Sarvabhutananda argues that a culture supportive of modern science was created in colonial India, which enabled independent India to continue investing in science and technology.

Professor K. Ramakrishna Rao's revised keynote address on "Science and Spirituality: Is There a Common Ground?" (Chapter 2) is the first full-length scholarly chapter in the book. One of India's leading psychologists, former vice-chancellor of Andhra University and twice chair of the Indian Council of Philosophical Research, Rao critiques "scientism," the dogmatic assertion of the supremacy of scientific knowledge, which he contends is a set of beliefs. Yet science is not merely an ideology but is an ongoing search for truth. It is not a settled body of knowledge. Nor is it a set of inviolable rules. With carefully chosen examples, Rao shows that what passes for science is often merely scientism; it is therefore necessary to question the dogmas of science, as it is the dogmas of religion. Contrasting the relationship between science and religion in the West with India, Rao argues that much of what we consider spirituality is also amenable to scientific inquiry. The chapter surveys some of the key issues and debates in the field of science and religion before asking for greater openness to truth from both sides.

Dr. R. P. Singh's "Dialectics of Culture and Science" (Chapter 3) examines the relationship between the two fields in the Western tradition. Singh, a professor of philosophy at Jawaharlal Nehru University, argues that science and religion are not opposed or separate but have been in a state of continuous tension and dialectical co-evolution in recent centuries. Singh's chapter in effect becomes a critique of modernity itself, engaging with the dominant Western tradition from Descartes to Weber, inclusive of Kant, Hegel, Marx, and Freud. In the end, he seems to suggest a return to the more holistic rationality of the Upanishads rather than the fragmentary reason of the Western Enlightenment thinkers.

In "Culture and Science in Twenty-first Century India" (Chapter 4), Dr. V. V. Raman discusses the aesthetic and cultural dimensions of culture and science in India. Raman, a noted mathematician, physicist, humanist, and professor of physics and humanities at the Rochester Institute of Technology, USA, critiques postmodernist interpretations of modern science, arguing that India's contribution has been, and will be, significant in the scientific understanding of experienced consciousness. Raman believes that while we should celebrate transrational beliefs about God and afterlife, we should also investigate analytically and with scholarly depth what aspects of ancient visions lead us to deeper understandings of ultimate and mind-related issues. At the same time, it is important not to negate naturalistic explanations of phenomena in the physical world offered by the meticulous methodology of modern science and modern mathematics. These "scientific" explanations transcend race and religion, geography and nationality; a growing number of people of Indic origin have been making significant contributions to them.

Fr. (Dr.) Job Kozhamthadam's chapter, "Modern Science in India and the Emergence of a New Worldview: Challenges and Opportunities" (Chapter 5), is a comprehensive survey of the face-off between science and religion in a globalizing world. Besides being an ordained Catholic priest, Kozhamthadam was the president of Jnana-Deepa Vidyapeeth, Pune, where he was also a professor of philosophy of science. He argues that in the past science was very much looked upon as a provider of important but dispensable amenities and comforts of life. However, today it is being recognized that science has and does play a far deeper and wider role in our world and society. Not only is modern science an integral constituent of our culture, it also is slowly but surely reshaping our culture, paving the way for the emergence of a new world order. India has always been the home of an ancient and rich culture, noted for its wide variety, diversity, and deep spirituality. Today, modern science is challenging to reshape India's deep-rooted values and time-tested customs. The chapter is a critical study of certain developments in modern science and technology which exert a profound influence on contemporary world culture in general and Indian culture in particular. Furthermore, it argues that India is in a privileged position to contribute substantially towards this dialogue since here both science and spirituality have always been taken very seriously.

The next chapter (Chapter 6), "Knowledge and Science in the Context of Indian Languages," is by Professor Probal Dasgupta, one of India's leading linguists and former professor at the Indian Statistical Institute, Kolkata. Dasgupta suggests that grammar was the foundational "science" in ancient India as geometry was to the ancient Greeks. If some of us begin to look at the approach to knowledge and science that developed around the notions of grammar in ancient Indian thinking, this transformation we undergo opens up for us the field of inquiry in ways that are not exclusively Paninian. His intervention contextualizes these debates in the context of generative reappropriation of ancient grammatical theorizing. Our standard access to ancient Indian grammatical work, he observes, is through contemporary linguistics and specifically through its grammatical theorizing. There is an imbalance here that needs to be corrected. Simply introducing a "socio-linguistic" tilt is not enough, for the implied binary between formal work and social inquiry is inappropriate. The generative reappropriation of ancient grammatical theorizing is a context that we need to overcome even as we respond to its imperatives. The social needs to be restored within grammatical inquiry itself; the attempt to introduce a sociological supplement misses the point.

Carrying on from Professor Dasgupta's Paninian analysis, Dr. Nirmalya Guha next looks at "Indian Deductive Systems: The Logical Basis for Indian Sciences" (Chapter 7). Guha argues that if logic is the study of the methods for distinguishing the incorrect arguments from the correct ones, then Indian deductive/reasoning systems are logical. We had several deductive systems in India, not just one. At least three schools of Indian philosophy, namely Nyāya, Mīmāmsā, and Bauddha, had their own deductive systems. All the Indian deductive systems are blockage based, a point which Guha's study tries to explain. But his central concern is with the Nyāya system as the representative of Indian logical tradition.

Dr. Madhumita Chattopadhyay, former professor and head of the Department of Philosophy at Jadavpur University, specializes in Buddhist philosophy, Buddhist logic, and contemporary Indian philosophy. Her chapter, "Motion Interpreted: A Bridge between Science and Spirituality" (Chapter 8), examines the concept of motion, which is one of the most important concepts in physics. The Buddhist logician Nāgārjuna, with the help of his dialectical method, concentrated on the metaphysical concepts of change and motion without attempting to deny the concepts such as "the space already travelled" (*gata*), "the space not yet travelled" (*agata*), or "the space being travelled at present" (*gamyamāna*). After a detailed analysis of each of these concepts from all possible alternatives, Nāgārjuna comes to the conclusion that an existing mover cannot exhibit motion; nor can a non-existent mover carry out the movement; nor can a person carry out the movement both existent and non-existent. Motion, therefore, is indescribable, ineffable, as are the Ultimate Truths of the Buddha. Nāgārjuna's attempt is to show that the analysis of even a scientific concept proves the truth of Buddha's teaching that language is not in itself an adequate means of describing the ultimate Reality. In short, even from a scientific concept we are led to the area of spirituality.

Dr. Sreekala M. Nair, former professor and head of the Department of Philosophy at the Sree Sankaracharya University, makes the important claim, similar to Probal Dasgupta's, in her chapter, "Indian Science and Semiotics: Some Reflections" (Chapter 9), that classical Indian thought put language and semiotics, not just logic or mathematics at the core of its notion of science. She observes, "While the West insists that science be logical, Indian thought structure places logic in science." Indian thought was supremely empirical rather than theoretical, with logic emerging out of it rather than individual facts or data being deduced from certain logical presuppositions. She argues that the latter methodology gave rise to modern science, but at a cost. Indian grammatical traditions offer another way, equally logical, to do science, a way whose potential has not, perhaps, been fully explored as yet. This forms the crucial epistemological difference between ancient Indian thought and modern science.

In the following chapter, "Integral Non-Dualism and Modern Science: Some Reflections" (Chapter 10), Professor Debabrata Sen Sharma offers an overview of four major schools of Indian non-dualism in their relationship to modern science. After considering the work of Mahayana Buddhists, Gaudapada and Sankara, and Ramanuja and Vallabha, he pays greater attention to Kashmir Shaivaism, which he calls "integral non-dualism." He argues that it is this view that comes closest to the unified field theory of modern physics.

Dr. Sampadananda Mishra, Sanskrit scholar at the Sri Aurobindo Society Pondicherry and Director, Vande Mataram Library, next explores the "Principles of Plant Taxonomy" (Chapter 11) to offer "A Fresh Insight into the Ancient Indian Methodology and Philosophy of Naming and Classifying the Medicinal Plants." Proper nomenclature and classification play important role in the systematization of any branch of knowledge. In this regard, Mishra demonstrates the extent to which the ancient Indian *rishis* and *acharyas* have shown transparency in their scientific observations. To them, to name was to touch the essence of the thing or object named. They could really enter into the soul or the consciousness of the thing or the object and then give the name as per their experience. We find a clear reflection of this in the names of the plants as they appear in various texts of Ayurveda. From the various names given to one plant, one can truly understand not only the various morphological characteristics of that plant but also the special medicinal properties that the plant has. It is still a mystery how the ancient Indian Vaidyas or medical scientists could discover the exact property of a plant and its multidimensional aspects when there was no such facility for empirical science as exists today. This chapter brings a fresh insight into this aspect and throws light on the ancient Indian methodology and philosophy of naming and classifying the medicinal plants.

Dr. Joy Sen, a trained architect and professor in the Department of Architecture & Planning, IIT Kharagpur, offers "A Scientific Approach to the Study of Indian Culture" (Chapter 12), discussing "Ecological Complementarity between Evolution and Involution." By this he means a holistic framework for conducting studies on culture and philosophy of science in India. This chapter addresses the need to redraw complementarities between scientific pursuits at a particular level of knowledge, with pursuits and inputs drawn from other levels such as culture, philosophy, and even normative sciences like psychology and bio-anthropology, which can extend to "deep ecological" concerns having moral, ethical, and "spiritual" dimensions. Sen's objective is to set indispensable iterations within the hierarchy to evolve a holistic framework of an emerging "new science."

Fr. (Dr.) Augustine Pamplany's chapter, "Reconciling Free Will and Determinism: An Indian Appropriation of Benjamin Libet's Neuroscientific Findings" (Chapter 13), addresses the question of moral responsibility from the Indian point of view, especially as it works in the domain of bioethics. Going beyond the traditional philosophical debate on free will and responsibility, which was dominated by the polarity between materialistic determinism and metaphysical dualism, the neuroscientific findings of Benjamin Libet concerning the nature of volition and free will have transformed the debate on moral responsibility into a scientifically tangible and controversial subject. Libet and his team inquired whether we are completely defined by deterministic laws of nature or have some independence in making choices and actions. An appropriation of Libet's data from the perspectives of the existing philosophical paradigms of moral responsibility suggests that Libet's findings are incompatible with libertarianism but seemingly fall in tune with compatibilism. However, a hermeneutically motivated analysis of Libet's data implies the scope of these findings beyond the compatibilist structures and suggests new directions for reconceiving the concept of free will and moral responsibility, founded mainly on some of the new ontological assumptions in quantum mechanics. Finally, an attempt is made to spell out some implications of this new framework for the Indian perspectives on free will and responsibility.

"Science and Spirituality from the Perspective of the *Mahabharata*" (Chapter 14), by Dr. Sitansu Chakravarti, is the next chapter. Referring to the *Mahabharata*, he tries to establish that the sustaining force for ethics is spirituality. The vast literature on logical positivism has a significant dearth of writings on ethics in so far as logical positivism deals with the realm of science. Although the ethical and aesthetic values find their limited inclusion eventually in the system, they are not contained there in the unified frame of the spiritual dimension of attitudinal virtues not to be accommodated in the science of physics. Science, he contends, is no substitute for spirituality, which is not (only) a search for the ontological state of the being but a pursuit for a value through the process of becoming. God is not a being in the Hindu system but is the value unconditional ananda, or joy, and is nothing but Existence, Consciousness, and Bliss, which are not marks of a being. The Gita says that all our actions are to be performed in the spiritual mode toward the attainment of the height of the spiritual state. Scientific pursuit must follow this route in line with the other pursuits in life without ever posing itself as a substitute for spirituality. However, the way science, quantum mechanics specifically, is relevant to spirituality is via its thrust on the unity of the universe, more so when one views things in the light of an all-pervasive consciousness. This thrust toward unity has a tendency to lead to the spiritual thrust of attaining harmony within and without, although the latter may not be reduced to the thrust toward unity to be found in science.

In his thoughtful and detailed review of concepts of health and disease, Dr. Bhushan Patwardhan, formerly professor and director of the Interdisciplinary School of Health Sciences, Savitribai Phule Pune University, now Vice-Chairman, University Grants Commission, argues for the urgent need to integrate modern and traditional systems of health. He asks for both sides to show mutual respect and openness so that they can collaborate rather than combat each other. In this context, the new stress on integrative medicine is heartening, especially in several areas such as oncology, cardiology, neurology, dermatology, psychiatry, and geriatrics.

The final chapter is called "A Connecticut Yankee in Indira's Court: A Brief Account of Modern Indian Science" (Chapter 16), by Dr. Raman Srinivasan. Originally a part of his PhD dissertation in the history of Indian science at the University of Pennsylvania, this chapter is a funny, moving, and deeply insightful narrative about how Indian science is uniquely culture specific, following patterns and structures that are embedded in Indian ways of seeing, thinking, and being. This does not make it any less "scientific," but its cultural context and contents are quite different from Western science. Srinivasan offers a fascinating account of Indian scientific achievements in fields as varied as nuclear technology and meteorology.

I am quite aware that each chapter has its own style, including that of translation from non-English sources such as Sanskrit, not to speak of a transliteration scheme which may not be identical. I have refrained from interfering too much in such matters. This is not necessarily the best practice but, in this case, expedient given that the quantum of such variance is not so significant as to render the concerned sections unintelligible to ordinary readers.

2

SCIENCE AND SPIRITUALITY
Is there a common ground?

K. Ramakrishna Rao

Science in its general sense is the search for truth. It is not a settled body of knowledge. Nor is it a set of inviolable rules. Rather, it is a continuous search leading to various degrees of certainty in our observations and understanding. If there are rules in doing science, they are guidelines to keep one on track. Statements of facts are approximations to and estimations of truth and not assertions of absolute certainty. From what we know of scientific research, we may form implicitly or explicitly a worldview, but to equate that worldview with science as such is to denigrate science into "scientism," a doctrinaire dogma. Scientism is, like spiritualism, a belief system. Spiritual refers to nonphysical aspects of being. It is overruled *a priori* by science in its doctrinaire sense and not in its true sense. It is possible to study spiritual phenomena scientifically. Psi research, which suggests a source of knowledge unaffected by physical complexities and unconstrained by space-time and sensory factors, provides legitimacy to scientific study of spirituality and gives reasons to question doctrinaire science that limits its study to physical phenomena. Inasmuch as parapsychologists do science in the spiritual area by studying non-physical phenomena, they appear to cover the common ground between science and spirituality.

In the Western tradition, science and spirituality are two disparate domains. Not only is there little in common between the two, they are also generally seen as based on and subscribing to diametrically opposed worldviews. In spite of this, in Western societies science and religion have somehow managed to coexist. The divinity school and the medical college, for example, situated side by side in the same campus, often teach contradictory things about who we are, where we come from, and what happens when we die. The chaplain and physician serve different functions based on opposing assumptions about human nature. The same person can be a scientist in the lab, following a materialistic worldview and working within a rationalistic framework, and then as a man of faith, devotedly offering prayers to the almighty God and seeking his blessings for his health and wellness. And surprisingly there is little experience of any psychological dissonance or cognitive conflict. In him, science and spirituality are stacked in two separate

10

compartments of his psyche. Science and spirituality thus coexist in the human condition. Why then is there an alleged conflict between them in academic discourse?

In the classical Indian tradition, the situation is entirely different. No inherent opposition is seen between the spiritual and the scientific world-views. Indeed, we find in it a magical synthesis of the two. Such a synthesis appears to be the trademark of the Indian ethos. In fact, I have argued elsewhere (Rao, "Magical Synthesis") that the method of magical synthesis constitutes the quintessence of Indianness, its manifest unity in diversity and the inherent diversity in unity the hallmarks of Hindu and Buddhist thought, respectively. How is such a synthesis possible? Can we sanitize science with spirituality and enrich spirituality with science?

In what follows, I will examine these possibilities and, in the process, attempt to present a brief account of the nature of science and what seem to be the main reasons for insulating science from spirituality. I will also point out what I consider to be priority strategies for bringing the two together. The latter includes the methods of research as well as the means of communication. I will argue (1) that the notion that there are objective criteria which distinguish genuine science in absolute and logically compelling terms from other knowledge claims is a pious myth, (2) that legitimacy in science is what we attribute to, rather than discover, in an area of study, (3) that science is a fascinating mixture of thought, action and passion, and (4) that a scientist's passion is no small determinant of legitimacy in science. I will then describe the current state of science-spirituality dialogue in terms of different perspectives of studying spirituality and their relevance to promoting the synergy between science and spirituality.

Legitimacy in science – is it a question of method?

As mentioned, current science is sharply contrasted with spirituality. The latter is considered the antithesis of science and is equated with mysticism, which is characterized as irrational and anti-scientific. Science itself is equated with a method of inquiry rather than a body of knowledge. If the essential aspect of science is its method, then an examination of its methods should settle the question of whether a given area of study is a legitimate science. Indeed, the belief in the existence of a uniform method underlying the practice of science in various disciplines was quite popular for a long time among those writing and reflecting on the nature of science. Perhaps the most widely known and influential statement of this viewpoint is by Karl Pearson in his *Grammar of Science*, first published in 1892. According to Pearson, science consists of a "classification of facts and the formation of absolute judgments" that are independent of the idiosyncrasies of the individual entertaining those judgments. The essence of science, according to Pearson, is its method and not the facts; and this method is the same in

11

all its branches. To quote Pearson: "The unity of all sciences consists alone in its method, not in its material" (15). The essential features of the scientific method are:

(a) Careful and accurate classification of facts and observation of their correlation and sequence; (b) the discovery of scientific laws by aid of the creative imagination; (c) self-criticism and the final touchstone of equal validity for all normally constituted minds.

(45)

Today few would subscribe to the notion that there is a single, objective scientific method by the pursuit of which we will be led indubitably closer to "truth." It has been pointed out that all attempts to precisely characterize the scientific method have so far failed to be convincing. J. Paul Feyerabend, for instance, has shown how even "the most advanced and sophisticated methodology" of science, such as the one described by Imre Lakatos, is inadequate in that there always exists a possibility that a research program that was once condemned as degenerative may be revived. Science as practiced by such celebrities such as Galileo is more ad hoc, and less methodical, than is generally presumed. Any description of the so-called scientific method can be shown to have been violated by at least one major advance in science. Therefore, Feyerabend concludes: "There is only *one* principle that can be defended under *all* circumstances and in *all* stages of human development. It is the principle: anything goes" (*Against Method* 14).

It is not only the anarchistic philosophers of science who have questioned the existence of the objective scientific method. James Conant, for example, writes:

There is no such thing as *the* scientific method. If there were, surely an examination of the history of physics, chemistry, and experimental biology would reveal it. . . . Yet, a careful examination of these subjects fails to reveal any *one* method by means of which the masters in these fields broke new ground.

(45)

Thus, it is difficult to argue for the existence of *the one* scientific method. But this does not necessarily invalidate the view that regards science as method: contemporary defense of this view can be found in Brown's *Perception, Theory and Commitment: The New Philosophy of Science* (1979).

What is this thing called science?

"Science," whatever it may mean, has had a profound influence on our lives, on our beliefs and actions. We think we know what we mean when we call someone a scientist and something scientific. All this, of course, does not necessarily imply that we all agree on what science really is. Nor is it the case that everyone would agree that science is the gateway to "truth." The views of philosophers of science vary all the way from reification of science to caricature and condemnation, from an absolute faith in the ultimacy of science as the only means of ascertaining facts and of advancing knowledge to the view that science is yet another ideology which has no special intrinsic certitude.

Whether or not the current high status enjoyed by science is justified on logical grounds, we are a party to the pyramiding of values that places science at the apex of the tools of inquiry leading us to truth. We want to be recognized as scientists because it is good and honourable to be so recognized as long as we are in the knowledge business. Methodological "scientism" deified as the sanctum of truth may be bad in principle. However, to reject science as a legitimate search for truth would be a serious regression to a primitive human condition.

It is widely believed that science is objective, that scientific knowledge is reliable and proven to be true, and that personal opinions and idiosyncratic speculations have no place in it. Closest to this common-sense view is the inductivist conception of science. Science, according to this view, starts with observations. Unprejudiced observations enable us to make statements about the world that are true or probably true. We are led from observation to generalized statements through the process of induction. But it has been pointed out that such a view is logically untenable. Inductive reasoning involves a leap from what is observed to what is not observed. There is no logical necessity that a conclusion reached by inductive reasoning is true even if the premises of inductive inferences are true. For example, one could conclude after making a large number of observations of swans in several parts of the world that swans are white. From this it does not logically follow that the next swan you observe will be white. David Hume showed over two hundred years ago that the attempt to establish the logical validity of induction is patently circular, arguing that, "what is possible can never be demonstrated to be false."

One way of solving this problem is to give up the inductive method of science altogether. This is what Karl Popper and his followers, who emphasized "falsification" instead of "verification," attempted to do. They concede that there is no logical necessity for scientific generalizations to be true. Science, according to them, is a set of hypotheses, hypotheses that are *falsifiable*. By *falsifiable* is meant the logical possibility of making an observation or set of observations that is inconsistent with the hypotheses. While no amount of witnessing white swans is logically sufficient to justify the conclusion that all swans are white, just one observation of a non-white swan is sufficient to falsify the statement that all swans are white. Singular statements of fact such as "this crow is black," however numerous, are insufficient to logically establish the truth of a universal statement such as "all crows are black."

Science, according to falsificationists, begins with problems. Problems lead to hypotheses. Hypotheses are subjected to test with an intent to falsify. Some will be falsified quickly, others may prove more successful. However, the process of falsification continues indefinitely. The theories that have withstood tests of falsification are not necessarily true but are superior to those that have failed. Science is an unending process of rejecting false hypotheses. The scientific worth of a theory is proportional to its degree of falsifiability. A theory that is clear and precise is more falsifiable than the one which is vague and ambiguous. Falsificationists much prefer "an attempt to solve an interesting problem by a bold conjecture, *even* (*and especially*) if it soon turns out to be false, to any recital of irrelevant truisms" (Popper 231).

A conjecture is bold if it is judged to be easily falsifiable. But such a judgment presupposes background knowledge. If, on the basis of available knowledge, a conjecture is unlikely to be proven, then its falsification is hardly an advance, but its confirmation, however, might constitute a major breakthrough. On the other hand, the falsification of a cautious hypothesis might be very significant, whereas its confirmation would be quite trivial.

It may be seen historically that a presumed falsification of a hypothesis did not always amount to its rejection. It is pointed out, for example, that "Newton's gravitational theory was falsified by observations of the moon's orbit." Bohr successfully persevered with his theory of the atom despite its early falsification. Chalmers illustrates the inadequacies of inductive as well as falsificationist accounts of science with reference to the Copernican revolution. At the time of the publication of Copernican theory in 1543, there were more things against it than in its favour. Without the development of the telescope and the new mechanics that eventually replaced Aristotle's, it would have been impossible to defend Copernican theory against those that it sought to replace. As Chalmers points out:

> New concepts of force and inertia did not come about as a result of careful observation and experiment. Nor did they come about through the falsification of bold conjectures and the continual replacement of one bold conjecture by another. Early formulations of the new theory, involving imperfectly formulated novel conceptions, were persevered with and developed in spite of apparent falsifications.
>
> (71)

The valiant attempts by Lakatos to improve on Popper with his emphasis on progressive research programs as opposed to degenerative programs is beset with similar problems. There are real difficulties in deciding whether one research program is better than the other. Again, programs that appeared to be degenerating at one time were revived at a later date and found to be fruitful. So, we have Feyerabend describing Lakatos' methodology as "a memorial to happier times when it was still thought possible to run a complex and often catastrophic business-like science by following a few simple 'rational rules' " (215).

Thomas Kuhn's notion of scientific paradigms is well known. What is important in Kuhn's characterization of the paradigms is that there are no easy criteria that determine the superiority of one paradigm over another. Inasmuch as rival paradigms subscribe to different metaphysical assumptions, no logically compelling demonstration of the superiority of one over the other is possible. The reasons for switching paradigms are more psychological and sociological than logical. Therefore, the arguments between those subscribing to rival paradigms are usually aimed at being psychologically persuasive rather than logically compelling.

From the foregoing it should be fairly obvious that it would be somewhat naïve to assume (1) that scientific inquiry is so objective that we can specify certain criteria that define genuine science and (2) that the generalizations of science are arrived at by truly objective observation. Our observations themselves are to a degree subjective. Scientific inquiry does not grow in a vacuum. It is carried out against the background of a culture with certain belief systems. These beliefs suggest problems as well as their probable resolutions. No science can claim absolute independence over its environment. To quote Ernest Schrödinger:

> The engaging of one's interest in a certain subject and in certain directions must necessarily be influenced by the environment, or what may be called the cultural milieu or the spirit of the age in which one lives. In all branches of our civilization there is one general world outlook dominant and there are numerous lines of activity which are attractive because they are the fashion of the age, whether in politics or in art or in science.
>
> (64)

The "internationality of science" or its apparent universal character is not an argument in favour of the objectivity of science. We have a similar consensus in international sports. It does not follow from it that these are the only possible ones. To quote Schrödinger again:

> In science we are acquainted only with a certain bulk of experimental results which is infinitesimally small compared with the results that might have been obtained from other experiments. . . . It would, generally speaking, be a vain endeavour on the part of some scientist to strain his imaginative vision toward initiating a line of research hitherto not thought of.
>
> (63)

15

Feyerabend put this somewhat differently, but more forcibly. The apparent universality in science, he argued, is due to "objectivization of the subjective" which enables the scientists to "transform their own personal or group idiosyncrasies into 'objective' criteria of excellence" ("Comments on 'Pathological Science: Towards a Proper Diagnosis and Remedy'" 53).

If science is a fashion, as Schrödinger acknowledged, it is also passion. We find a convincing exponent of this view in Michael Polanyi. Polanyi distinguished between three kinds of passion: first is the intellectual passion, which affirms the scientific value of certain facts; then, the heuristic passion, which provides the impetus for originality and creativity; and, finally, the persuasive passion, which is behind most controversies in science. "I certainly affirm," writes Polanyi, "that passion and controversy moved by passion, must continue in science and that a comprehensive revision of our philosophy of science is needed to give due weight to this essential aspect of scientific truth" (103).

There are thus severe difficulties in characterizing science as this or that. Yet the situation is not as hopeless as some anarchistic philosophers would picture it. There are some basic assumptions on which most of us who call ourselves scientists can agree. We would agree, I think, that there is a world out there which is real and relatively independent of us. That world can be known through observation and experiment. Despite certain subjective characteristics of experience, most of us experience the outside world in similar ways. While the principle of induction and the notion of the uniformity of nature may not be logically compelling, they seem to work pretty well in practice. It does not follow, however, that what we see out there and what we experience through our sensory channels are the only real things. Our problems in understanding science are at least in part due to our failure to appreciate its complexity. Science is a complex activity carried on against a certain background by men and women of flesh and blood. Therefore, proper understanding is possible only when we consider the business called science in the light of the beliefs and behaviour of those engaged in that trade.

It seems to me that science is a complex milieu consisting of scientists and their thoughts, actions, and passions. Thought is a scientist's background knowledge, which suggests problems and possible solutions. Action is the method, which prescribes how questions should be posed and treated and how to verify initial assumptions in relation to the questions raised. Passion is that which is involved in a scientist's mode of discourse and his interpretations of the results. It is what colours his statements and meanings, which he relates to truth and falsity. These three elements – thought, action, and passion – blend in any given scientific enterprise to give us a mix called science. Inasmuch as the proportions of these vary from area to area and inquiry to inquiry, we have sciences of various shades and persuasions.

Now within this framework, let us examine the question of legitimacy in science and whether we can have a science that addresses issues relating to spirituality. Surely, anything does not go. At the same time, there is no perfect knowledge and no pure method in science. The methods in use are no more sacrosanct and infallible than the knowledge we have of the world. At a given time, however, we have a body of knowledge, which we have little reason to question. If nothing else, it seems to work pretty well in practice. Again, a certain kind of activity appears to generally characterize those who aspire to become scientists. Thus there is a general consensus on these points. But the very essence of science is not perpetuation of the status quo but change and advance. The scientist is driven by a passion to be creative and original so that he can break new ground. Therefore deviations from the practices he has learned and beliefs he has entertained are a part of the scientific process. Normally such deviations cause no concern; they are even encouraged. But when a scientist makes a claim and his results are interpreted to constitute a threat to the integrity of certain "established" claims in science, all sorts of efforts to attribute non-legitimacy to that scientist's work will be made. This is the reason why few scientists venture into the investigation of phenomena considered spiritual.

In dealing with claims that conflict with current "confirmed" knowledge, the first attempts will involve the examination of the accuracy of the data and the source of error within a frame of reference that the critic as well as the claimant shares as scientists. This can be done with relative objectivity. Once this phase is passed, the controversy will tend to be essentially rhetorical, since a deviant claim cannot be dismissed on logical grounds. The most effective, though not valid, means of rejection is to deny legitimacy on the grounds of non-conformity with the "established laws" and approved methods. The fact that the so-called laws are not infallible and that the methodological rules have been violated in the past by those who have carried out legitimate science makes it very clear that legitimacy is what we attribute to, rather than what we discover in, a scientific claim. It becomes clearer when we realize that a lack of methodological rigor is seldom considered sufficient to determine legitimacy. The questions of legitimacy arise at the level of passion but are promoted by the perceptions of their incompatibility with our background thought and action. Thus the effort to establish the legitimacy of scientific claims is often directed at appealing to the passionate nature of scientists rather than at demonstrating its intrinsic legitimacy on logical grounds.

Limitations of current practices of science

If science is essentially a method, it follows that there is no single worldview that is sacrosanct and that no subject can be ruled out of court for scientific research on *a priori* grounds. It is here the current practice of science stretches itself to champion a worldview that rules spirituality out of the bounds of science. The stretching is both methodological and doctrinaire.

17

From the methodological side it is held that the scientific method is applicable in certain areas and not in others. Spirituality is one of those areas that are inherently resistant to scientific investigation, and therefore it is an inappropriate subject matter for science. This argument makes two basic assumptions. First, it is assumed that there are clear-cut methods of scientific investigation that unambiguously identify scientific inquiry as opposed to other quests. Second, it is presumed that inquiry in science depends on the availability of data that are shared and objectively available. We have pointed out in the previous section why the notion that there is a single set of rules for doing science is suspect and how the method of inquiry in science appears to shift in the hands of scientists to suit the select area of study. Again, it is an unwarranted prerequisite that the scientific data be objective, third-person observations. There is no intrinsic reason why first-person experience cannot be the datum for scientific inquiry.

It is true that the tremendous advances in physical sciences and technologies based on them during the past century or so are indeed time-tested testimonies for the efficacy of third-person observation as a valuable tool for doing science. Indeed, it is regarded with some justification that it is the pivotal point, the fulcrum for the turnings of science. However, it could also be argued that the relative lag in the progress of human sciences is also a consequence of the emphasis of third-person observation to the relative neglect of experience, which is the mainstay for human science. Thus, while the emphasis on third-person observation appears appropriate in the domain of physical sciences, its indiscriminate extension to all areas, especially human sciences, is not warranted.

The current practice of science goes beyond its use as a method to include a worldview that may be characterized as essential materialism, that all the things and events in the universe are manifestation of matter driven by physical force and that in the final analysis all events are reducible to physical states. This is what Charles Tart labels as "scientism" – "a dogmatic, psychological hardening of materialistic belief systems with emotional attachments, rather than authentic science." Tart argues that the degeneration of science into scientism has the effect of "wholesale undermining of spirituality by orthodox science." This, Tart argues, "is not only unhealthy but scientifically, *factually* wrong" [emphasis in the original] (22).

There is another aspect of science that needs a mention here. Science is generally considered value-free. In other words, a scientist qua scientist has no ethical stake in doing science. He discovers facts and invents tools without regard to their potential value to promote or undermine common good. This is left to others to handle. A scientist is supposed to work in an ethical vacuum. So we have nuclear technologies and stem-cell research, which could be used to destroy or distort life on the planet. Values are associated with religion and spirituality. The divorce of science from spirituality is at the root of the notion of value-free science. There is thus good reason for sanitizing science with a dose of spirituality. To make this possible we need to discover the common ground between science and spirituality.

While most scientists are content with the uneasy coexistence of science and spirituality and generally tend to ignore the manifest conflict between the two opposed worldviews, there are some brave ones who broach the subject of reconciliation between the two. These efforts fall into two broad categories. One is in the direction of secularizing the sacred; the other takes head-on the presumed opposition of the two by collecting data that fundamentally challenges the view that humans are essentially "materialist animals" and electrochemical machines, and suggests a spiritual view of the world. The latter rejects the notion that science is the antithesis of spirituality and that the reconciliation of the worldviews of science and spirituality is possible at the foundational level. The former, epidemiological studies of spirituality, though promoted by spiritually inclined scientists, is no more than finding a rational place for spirituality in life, as we will presently see.

Epidemiological studies of spirituality

Recent epidemiological research in the area of religiosity and a variety of health conditions suggests significant relationship between religious practices taken as indices of spirituality and human health (Koenig et al. 2001). In a series of studies by H. G. Koenig and colleagues, various aspects of the relationship between religious activity and health outcomes were looked into. They examined the effect of religious involvement on mortality, which showed a significant association between private religious activity such as prayer and longer survival in certain population groups (Helm et al., "Effects of Private Religious Activity on Mortality of Elderly Disabled and Nondisabled Adults"). They also found significantly less anxiety disorders among the people who attended religious services regularly (Koenig et al., "Religion and Anxiety Disorder") and an inverse relation between religious coping and scores on depression scales (Koenig et al., "Religious Coping and Depression in Elderly Hospitalized Medically Ill Men"). Several studies with different groups ranging from terminally ill (Reed 1986, 1987) to healthy adults (Mattlin et al.) suggest that a significant majority of people report that they use religion as a coping mechanism to deal with health problems and other stressful situations. Steffen, Hinderliter, Blumenthal, and Sherwood reported that among African Americans, religious coping is associated with reduced blood pressure. As Koenig, McCullough, and Larson observe, among the sixteen studies that have examined possible relationship between religious involvement and blood pressure, fourteen report lower blood pressure among the more religious.

Goldbourt et al., reported that in a twenty-three-year follow-up study, the risk of death from coronary artery disease (CAD) was 20% lower among the more orthodox Jews than the less orthodox or non-believers. In their

19

study of religious struggle as a predictor of mortality among medically ill elderly patients, Pargament, Koenig, Tarakeshwarm and Hahn found that the patients who experience religious struggle in comparison to those who religiously cope with their illness are at a greater risk of death. In a six-year follow-up study of 3,968 older adults, they observed those who attended religious services at least once a week appeared to have a survival advantage over those who attended services less frequently. This effect of religious activity on survival, they contend, is equivalent to that of non-smoking vs. smoking on mortality (Koenig et al., "Does Religious Attendance Prolong Survival?").

There are several significant studies that explored the relationship between religiosity and a variety of health conditions. In about 150 studies on alcohol and drug abuse and religious involvement, most of the studies "suggest less substance abuse and drug abuse and more successful rehabilitation among the more religious" (Koenig et al., *Handbook of Religion and Health*). Also, numerous studies investigated the effect of religion on mental health, delinquency, depression, heart disease, immune system dysfunction, cancer, and physical disability. (For a comprehensive review of research in these areas, *see* Koenig et al. 2001).

Surveys of literature and meta-analysis of published research seem to confirm the claims of individual researchers linking religious practices with better health outcomes. For example, in a systematic and comprehensive review, Townsend, Kladder, Ayele, and Mulligan assessed the impact of religion on health outcomes. They reviewed all experiments involving randomized controlled trials published between 1996 and 1999 that assessed the relationship between religious practices and measurable health variables. The review revealed that "religious involvement and spirituality are associated with better health outcomes, including greater longevity, coping skills, and health related quality of life and less anxiety." In a meta-analytic review of twenty-nine independent samples, McCullough et al. (2000) reported that religious involvement has a strong positive influence on survival.

If religious involvement does have beneficial health outcomes, as many of the published reports in the West seem to suggest, then we may ask: How does this relationship work? What is its modus operandi, the process that underlies the presumed effect? What is the channel? Who is the source? What is the relevance of spiritual phenomena in this context? These important, though often tricky, questions have no easy answers. The favoured explanation is a secular one. Religious beliefs and practices may have psychological effects which in turn bring about somatic changes. If indeed religious beliefs and activities help to reduce anxiety, stress, and depression, they could also help to shield their negative effects on general health and well-being.

Koenig, Larson, and Larson observe that when people become physically ill, many rely heavily on religious beliefs and practices to relieve stress, retain a sense of control, and maintain hope and a sense of meaning and purpose in life. They suggest that religion (1) acts as a social support system, (2) reduces the sense of loss of control and helplessness, (3) provides a cognitive framework that reduces suffering and enhances self-esteem, (4) gives confidence that one, with the help of God, could influence the health condition, and (5) creates a mindset that enables the patient to relax and allow the body to heal itself. Again, the values engendered by religious involvement, such as love, compassion, charity, benevolence, and altruism, may help to successfully cope with debilitating anxiety, stress, and depression.

Thus we find a reassuring secular interpretation of spirituality, which interestingly suggests the significance of being spiritual in one's life without addressing the basic question whether there is any scientific validity for spiritual phenomena. It merely provides for legitimacy of religious lifestyle without legitimizing spirituality itself as genuine. In some ways the sidestepping of the basic issue may be comforting for the scientist engaged in spiritual practices. At best it may be considered a healthy rationalization for avoiding intellectual dissonance but hardly a resolution of the problem. In fact, there are some published epidemiological studies that raise some unsettling questions about the adequacy of the secular interpretation of the benefits of spirituality. Let us consider briefly the research on remote intercessory prayer and its effect on recovery from illness.

In his 2004 book *The Case of Distant (Remote) Intercessory Prayer*, Michael Miovic refers to two well-documented cases of spiritual healing by a Russian healer, Nicolai Levashov, reported by Koopman and Blasband ("Two Case Reports of Distant Healing"). In one case, a baby girl was completely healed from glioblastoma multiforme (GBM), considered the most aggressive form of brain cancer, believed to be incurable, and ultimately fatal. In another case, the same healer is reported to have successfully cured a boy who was diagnosed with testicular absence at the age of one month. At the age of eleven, serial tests of free testosterone showed near absence of any hormone production. Then, in 1999, Levashov began distant healing on the boy. By August 2000, testosterone reached near normal levels. By 2002, doctors reviewing the case "acknowledged that functional testicles had appeared in a genetic male who had presented well past the age at which testicles can develop" (Miovic 129).

A number of studies gave positive evidence linking intercessory prayer with beneficial health outcomes. Intercessory prayer involves praying for others' benefit. In some of these studies, the patients did not know that someone was praying for them. Yet their condition seemed to have improved compared to the control group of patients who did not have the benefit of someone praying for them. In a double-blind study involving 393 coronary care patients, Randolph Byrd (1988) divided his subjects into two randomized groups. One group was the intercessory prayer group and the other was the control group. Neither the physicians attending on them nor the patients themselves knew which patients were being prayed for. Also, those who actually offered prayers did not know the patients for whose recovery they were praying. Results showed that the patients in the intercessory prayer group experienced significantly fewer episodes of congestive heart failures, experienced fewer cardiac arrests, received fewer antibiotics, and required less respirator support and medication. Byrd's 1988 study was criticized for "multiple analyses" and "methodological crud factor" by Chibnall et al. ("Experiments in Distant Intercessory Prayer"). However, J. E. Kennedy, on re-evaluating the data, concluded that, "the results for two of the outcome measures are significant at the .05 level even after conservatively correcting for 29 multiple analyses" ("Commentary on 'Experiments on Distant Intercessory Prayer,'" 181). Harris et al. (1999) conducted a double-blind study of distant intercessory prayer with 990 patients in the cardiac care unit. In this study with randomized controlled trials, it was observed that the experimental group (the prayed for patients) recovered better than the control group of patients. The results are statistically significant, even after correction for multiple analyses.

In a meta-analysis of published studies, Mueller, Plevak, and Rummans found that randomized controlled trials had shown a significant positive effect between intercessory prayer and recovery from coronary disease. They observed that addressing the spiritual needs of the patient may enhance recovery from illness.

Recent research on the effects of intercessory prayer on human health seems to yield inconsistent results, casting doubt on the previous findings. For example, Benson et al. (2006) carried out an extensive study involving patients at six US hospitals and on six hundred patients undergoing coronary bypass graft (CABG) surgery. The patients were randomly assigned to one of three groups. In Group 1, the patients were informed that they may or may not be prayed for, but they were prayed for. In Group 2 like Group 1, the patients were also uncertain whether they would be prayed for. However, they were not prayed for. In Group 3, the patients were told that they would be prayed for, and they did indeed receive intercessory prayer. We can clearly see that Group 3 is the target group where the intercessory prayer effort is most likely to be present.

The results of the study were analyzed in terms of uncomplicated recovery after surgery. The most striking finding is that 63% of the patients on Group 3 as per the interim analysis has a complication after surgery compared to 51% in Groups 1 and 2. The difference is statistically significant (p= .003). The final analysis also showed similar results. In Group 3, 59% of the patients had complications after surgery compared to 52% in Group 1. The difference between the two groups is statistically significant. Other secondary analysis also showed that "patients in Group 3 were consistently more likely to have a complication than those in Group 1 across the planned subgroup analyses." Notwithstanding the above significant differences between groups, Benson et al. conclude: "Intercessory prayer itself had no effect on complication-free recovery from CABG, but certainty of receiving intercessory prayer was associated with a higher incidence of complications."

Krucoff, Crater, and Lee critique the study design of the Benson et al. (2006) on the following grounds: (1) the use of just a few prayer groups may have affected the actual prayers performed, (2) the reason for use of a certainty vs. uncertainty model instead of a double-blind method have been left unexplored by the authors, (3) the certainty/uncertainty design could have provided insight into the placebo effect, which was also not explored, and (4) the authors have not discussed the significant results that indicated a worsening of the condition beyond explaining them as a chance finding. Krucoff et al. comment,

> If the results had shown benefit rather than harm, would we have read the investigators' conclusion that this effect "may have been a chance finding," with absolutely no other comments, insight, or even speculation?" Krucoff et al. further state that "the study results appear to reflect more the cultural bias that healing prayer could only seriously be explored for effectiveness, not for safety issues. Culturally, 'harm' resulting from prayer is generally ascribed to overtly 'negative' prayer, such as hateful prayer, voodoo, spells, or other black magic. Positively intended intercessory prayer is considered *a priori* to be only capable of doing good, if it does anything at all.
>
> (763)

Clearly, the results of research on the effects of intercessory prayer on health are inconclusive. However, research on intercessory prayer described above does not stand alone. There is a massive database for rejecting essential materialism, the doctrinal twin of methodological scientism, carefully collected over a century of effort. I refer to psychical research, or parapsychology at it has come to be known, which offers substantial evidence in support of a worldview that accommodates the positive findings of intercessory prayer on health outcomes.

On parapsychology

Since the beginning of the recorded history of humankind, there have been reports suggestive of paranormal communication (for further reading on this, please refer to Inglis' *Natural and Supernatural: A History of the Paranormal from Earliest Times to 1914*). There is growing scientific evidence to suggest the possibility of acquiring information that is apparently received independently of our sensory process, as in extrasensory perception (ESP), and of direct action of mind over matter independent of our motor system, as in psychokinesis (PK) (Rao and Palmer, "The Anomaly Called PSI: Recent Research and Criticism"). Also, there are a number of cases of a person claiming to remember events in a previous life (Stevenson 1974, 1977). Clearly, all these phenomena are inconsistent with the Western scientific worldview. Therefore, they are regarded as anomalies in need of explanation.

From the time of J. B. Rhine's monograph *Extra-Sensory Perception*, first published in 1934, literally hundreds of carefully carried out experimental studies have accumulated a massive database that not only makes a strong case for the existence of psi (a term that includes ESP and PK) but also offers interesting insights into the nature of psi. For example, the results of psi research show that the physical aspects of the target, such as size, shape, colour, and form, do not seem to have any intrinsic effect on psi. Neither do space and time or the causative complexity of the psi task. Any hypothetical relationship of distance to ESP must assume that there is some energy transmission between the subject and targets that is inhibited by the distance factor. But if precognition is a fact, and we have strong evidence to believe that it is, what is the nature of this transmission that occurs between the subject and the not-yet-existing target? Thus, the evidence for precognition and the success of ESP experiments over long distances lead one to believe that space and time are not constraining variables as far as psi is concerned. Another significant aspect of psi is the relative ineffectiveness of task complexity in constraining psi. Stanford has reviewed the relevant literature and concluded that "the efficiency of PK function is not reduced by increases in the complexity of the target system" (375).

If psi is unconstrained by space and time and the complexity of the task, and if the psi situation is such that distinctions between thought and matter, cognition and action, and subject and object become less than meaningful, it would seem that psi may function beyond the familiar categories of understanding and may point to a state of being that cannot be properly classed as mind or matter. Psi phenomena raises the question whether there exists a realm of reality beyond the phenomenal world of appearance, which is primarily a product of our information-processing capabilities and mechanisms. One may rightly wonder whether we are not dealing here with the Kantian "thing in itself." What is interesting, however, is that the thing in itself which, according to Kant, must remain forever beyond the human reach may in fact be the reality to which psi has direct access, a reality assumed by most religions. Thus para-psychological phenomena may be pointing to another source and mode of knowing that renders the spiritual worldview plausible. Interestingly, the source is suggested by data accumulated by following rigorous scientific methods. Thus, psi research may suggest the common ground between science and spirituality. In fact, psi research provides strong evidence that human physiological states can be influenced by para-normal means. These results give rudimentary support to the epidemiological research on the effects of spirituality on health.

There is a large empirical database accumulated over the years by William Braud and associates that provides strong evidence suggestive of the possibility of influencing the physiology of a remotely situated person by sheer mental intention of another person. Braud and Schlitz review eight separate experiments in which the subjects attempted to influence remote biological systems by simply wishing such a change. The crucial difference between prayer and such wishing is that no supernatural being is invoked in the wish phenomenon, unlike in the prayer, which is generally directed at seeking the help of God to grant the wish. The results of the experiments by Braud and associates show that a subject by mental intention alone could influence in the desired direction (1) the autonomic nervous system activity of a remotely situated person, (2) the muscular tremor and ideo-motor reactions, (3) mental imagery of another person, and (4) the rate of haemolysis of human red blood cells in vitro. There is no reference in these studies to supernatural beings or non-testable entities. As Braud points out, based on the overall statistical results, the distant mental influence effects are relatively reliable and robust. The magnitude of the effects is not trivial and is comparable to self-regulation effects. The ability to mentally influence is apparently widely distributed.

Parapsychology and spirituality

J. B. Rhine, the founder of experimental parapsychology, who helped to narrow the focus of the field to empirically researchable areas, felt that the parapsychology of religion is an important art to pursue. Some fifty years ago he discussed the relevance of parapsychology to religion in an editorial in the *Journal of Parapsychology*. "If para-psychology finds answers to the questions for which religious doctrines have been developed in the past," he wrote, "there is no reason why these should not replace the earlier conceptions in much the same way that chemistry has replaced alchemy and scientific medicine has taken the place of the practices of pre-scientific days" (Rhine, "Telepathy and Clairvoyance Reconsidered."1).

Rhine proposed to the FRNM Board of Directors in 1968 that they consider establishing a separate branch of the FRNM devoted to the parapsychology of religion. Recognizing that "organized religion is not yet ready for a deliberate step on this firmer pavement of scholarly testing," Rhine felt that "the time has come when a quiet research program could be undertaken by a branch of the FRNM which we might call the Institute for the Parapsychology of Religion."

Dr. Rhine's interest in the parapsychology of religion is of course a continuation of such interest among a long line of psychical researches. The prominent among them are F. W. H. Myers and William James. Myers' hunch that the subliminal consciousness is the spiritual component of our personality, which is not only the source of paranormal awareness but is

25

also the one that survives bodily death, adds a distinctive new dimension to the Western approaches to the study of consciousness. Similarly, the studies of religious experience by William James and his understanding of mystic states are a lasting contribution to the parapsychology of religion.

But Rhine was on firmer ground than his predecessors. Rhine saw in his results the scientific basis for believing that there is "an extra-physical element in man" and for rejecting the philosophy of materialism, "the chief enemy of religion." He wrote in *New World of the Mind* (1953):

> The conclusion is inescapable that there is something operative in man that transcends the laws of matter and, therefore, by definition, a non-physical or spiritual law is made manifest. The universe, therefore, does not conform to the prevailing materialistic concept. It is one about which it is possible to be religious; possible, at least, if the minimal requirement of religion be a philosophy of man's place in the universe based on the operation of spiritual forces.
>
> (227)

The basis for Rhine's belief that para-psychological phenomena point to "extra-physical" or "non-physical" reality is the finding that psi is unconstrained by space-time barriers. Once the physical barrier is broken, Rhine thought, new possibilities open up for understanding the spiritual powers of man and the possibility that the spirit survives bodily death. Rhine also was impressed by the similarity between religious forms, such as prophecy and prayer, and types of psi, such as precognition and telepathy (Rhine, "Comments: Second Report on a Case of Experimenter Fraud").

The Eastern traditions, especially the Hindu and the Buddhist, recognize that psychic (paranormal) abilities are a natural outcome as one progresses on the path to achieve states of transcendence and has religious experience. They warn, however, that these should not become ends in themselves but be taken as guideposts to recognize one's success in the pursuit of transcendence. Therefore we read in *Visuddhimagga*:

> Psychic powers are those of an average man. Like a child lying on its back and like tender corn it is difficult to manage. It is broken by the slightest thing. It is an impediment to insight, but not to concentration because it ought to be obtained when concentration is obtained.
>
> (Buddhaghosa 113)

Parapsychology has substantial implications for religion, constituting a new area of study, the parapsychology of religion. The parapsychology of religion can (1) provide descriptive accounts of religious experience that would apply universally, independent of individual religions; (2) construct empirical tests for validating specific forms of religious experience; (3) discover significant correlations between religious beliefs and practices on the one hand and psychic abilities on the other; (4) formulate instructional aids for those desiring to have religious experience; (5) give training in therapeutic and counseling skills to clergy so that they can responsibly deal with those having religious or pseudo-religious experiences; and (6) help bridge the chasm between religion and science.

All major religions are based on the teachings of a few, described variously as prophets, saints, *rishis*, and so on. All these teachings are claimed to be based on their intense personal experiences. What is the nature of these experiences qua experiences? I am persuaded that a systematic and descriptive study of such experiences would reveal something more basic to religious experiences than do the content of these experiences translated to us in common language forms. Religious experience, we are told, is essentially ineffable. The differences in the statement of the content of religious experience as described in various religions may be due to the difficulties of translating the ineffable experience into cognitive content. Religious experience, the mystic experience, the peak experience, and all paranormal experiences may have one thing in common. They are the encounters with consciousness *as such*, pure consciousness in which there is no subject-object distinction, no content but a transformational process that often results in remarkable behavioral changes and beliefs and sometimes translates itself into informational content.

If phenomenological studies of those who have religious and paranormal experiences give us an appropriate description of the essential aspects of such experience, it should be possible to construct empirical tests for validating those experiences as involving genuine access to consciousness as such. For example, criteria for identifying religious experience and measuring its depth and intensity may be developed. Such criteria would be useful in distinguishing genuine religious experience from pseudo-religious experience in a manner similar to the one of distinguishing the fake psychic experiences from the genuine (Rao, "Parapsychology: Implications for Religion and Science").

If, as William James states, "the mother sea and fountain-head of all religions lie in the mystical experiences of the individual" (Allen 425), and if mystical experiences may be had by following certain procedures such as meditation, it should be possible by a systematic study of these procedures and practices as well as other psychic development strategies to develop instructional aids for those aspiring to have religious experience. In the Eastern traditions, as in Patanjali's Yoga, there exists elaborate descriptive accounts of recommended practices that are believed to lead one to experience higher levels of consciousness, including religious experience. These deserve to be studied carefully and systematized appropriately based on empirical fact.

27

The parapsychology of religion so pursued could be invaluable to clergy in their training not only for their own pursuit of personal religious growth but also for providing counseling to the clients in their congregations. As in the realm of psychic phenomena, there are pseudo-religious phenomena that mimic the genuine. It is therefore necessary to understand genuine religious experiences as distinct from pseudo, religious-like experiences. The latter may indicate maladjustment and mental illness. The emphasis on the importance of teacher, the *guru*, in the Eastern religious traditions is relevant here. The *guru* is the guide who at every stage of religious development is able to counsel the disciple to proceed in the right direction, avoiding the pitfalls (Rao, "Psychology of Transcendence: A Study in Early Buddhistic Psychology").

We need no longer base the validity of our religious beliefs and practices on doctrinal philosophies and theologies inasmuch as they are born out of actual experience and their veridicality can be attested by observation. All religions claim that their "truths" are received as revelations, which are essentially extra-sensory forms of knowledge. That knowledge can be obtained independently of rational discourse and sensory experience is a fact that parapsychology has sought to establish. The parapsychology of religion may thus provide the empirical ground for dealing with religious beliefs and practices and thus bridge the chasm between religion and science.

Research has already revealed several commonalities between religious experience and psi phenomena. They point to the possibility of some of our experiences being non-local and trans-personal. Factors such as belief and motivation seem necessary for manifesting psi as well as religious experience. Internal attention states such as meditation and conditions of sensory deprivation are known to facilitate the occurrence of religious experience as well as psychic phenomena. Finally, religious experience has important transformational consequences for the one who has the experience. Paranormal phenomena also seem to have similar effects. People with near-death experiences report significant lifestyle changes after the NDE. Abraham Maslow also reports such transformations when people have what he calls peak experiences. Therefore, one wonders whether in all these cases people are not accessing consciousness as such.

Inasmuch as the characteristics of psi are inconsistent with the materialistic conception of the universe and point to a reality closer to the one that underlies major religious traditions, we may expect parapsychology to provide the necessary interface between science and religion. To be sure, science shall remain science (i.e., rational inquiry into reality), and religion shall remain religion (i.e., revelatory experience of reality). But they do not have to contradict or negate each other.

In the above, we did not distinguish between religion and spirituality, but it is assumed that spirituality is what lies behind religious practices. This is an assumption that may not be justified at the ground level when we look at some of the practices that go in the name of religion. Therefore, perhaps it is best that we replace religion with spiritual. Also, we should note that spiritual is not the same as spiritualism. The latter is doctrinaire like scientism.

I made a passing reference earlier to sanitizing science, which means filling up its moral vacuum. Science and scientists cannot abdicate their moral responsibilities. In the human condition there is no value-empty endeavour. Science needs to deal with the matter of morals, as it should with spirituality. Values cannot be anchored in essential materialism or spiritual transcendence. Genuine morals are not cognitively constructed out of rational discourse. Nor are they realized in transcendental loneliness. I believe they originate in the common ground shared by science and spirituality, a trance-cognitive state suggested by psi research and which is implied in a spiritual worldview. As Tart so eloquently argued, we are in some ways electrochemical machines, often acting like materialist animals, but we are also spiritual beings with enormous potential. The person is thus the agent of science as well as spirituality. Para-psychological phenomena prominently display the common ground covered by science and spirituality. In a meaningful sense, it may be seen as a spiritual science.

Works cited

Allen, G.W. *William James*. New York: Viking, 1967.

Benson, H., J.A. Dusek, J.B. Sherwood, P. Lam, C.F. Bethea, W. Carpenter, S. Levitsky, P. Hill, D.W. Jr Clem, M.K. Jain, D. Drumel, S.L. Kopecky, P.S. Mueller, D. Marek, S. Rollins, and P.L. Hibberd. "Study of the Therapeutic Effects of Intercessory Prayer (step) in Cardiac Bypass Patients – A Multi-Center Randomized Trial of Uncertainty and Certainty of Receiving Intercessory Prayer." *American Heart Journal*. Vol.15 (2006): 934–942.

Braud, W.G., and M.J. Schlitz, "Consciousness Interactions with Remote Biological Systems: Anomalous Intentionality Effects." *Subtle Energies*. Vol.2.I (1991): I–46.

Brown, H.I. *Perception, Theory and Commitment: The New Philosophy of Science*. Chicago: University of Chicago Press, 1979.

Buddhagosha. *Visuddhimagga [The Path of Purity]*, 3 Vols. Trans. M. Tim. London: Oxford University Press, 1923.

Byrd, R.C. "Positive Therapeutic Effects of Intercessory Prayer in a Coronary Care Unit Population." *Southern Medical Journal*. Vol.81 (1988): 826–829.

Chalmers, A.F. *What Is This Thing Called Science: An Assessment of the Nature and Status of Science and Its Methods*. Milton Keynes: Open University Press, 1978.

Chibnall, J.T., J.M. Jeral, and M.A. Cerullo. "Experiments in Distant Intercessory Prayer: God, Science, and the Lesson of Massah." *Archives of Internal Medicine*. Vol.161 (2001): 2529–2536.

Conant, J.B. *Science and Common Sense*. New Haven: Yale University Press, 1951.

Feyerabend, P.K. *Against Method: Outline of an Anarchistic Theory of Knowledge*. London: New Left Books, 1975.

———. Comments on "Pathological Science: Towards a Proper Diagnosis and Remedy." *Zetetic Scholar*. Vol.6 (1980): 52–54.

Goldbourt, U., S. Yaari, and J.H. Medalie. "Factors Predictive of Long-Term Coronary Heart Disease Mortality Among 10,059 Male Israeli Civil Servants and Municipal Employees." *Cardiology.* Vol.82 (1993): 100–121.

Harris, W.S., M. Gowda, and J.W. Kolb. "A Randomized, Controlled Trial of the Effects of Remote, Intercessory Prayer on Outcomes in Patients Admitted to the Coronary Care Unit." *Archives of Internal Medicine.* Vol.24 (1999): 79–88.

Helm, H., J.C. Hays, E. Flint, H.G. Koenig, and D.G. Blazer. "Effects of Private Religious Activity on Mortality of Elderly Disabled and Nondisabled Adults." *Journal of Gerontology: Medical Sciences.* Vol.55A (2000): M400–M405.

Hume, D. *Treatise on Human Nature.* London: Dent, 1939.

Inglis, B. *Natural and Supernatural: A History of the Paranormal from Earliest Times to 1914.* London: Hodder and Stoughton, 1977.

Kennedy, J.E. "Commentary on 'Experiments on Distant Intercessory Prayer'." *Archives of Internal Medicine: Journal of Parapsychology.* Vol.66 (2002): 177–182.

Koenig, H.G., H.J. Cohen, D.G. Blazer, C. Pieper, K.G. Meador, F. Shelp, V. Goli, and R. DiPasquale. "Religious Coping and Depression in Elderly Hospitalized Medically Ill Men." *American Journal of Psychiatry.* Vol.49 (1992): 1693–1700.

Koenig, H.G., S. Ford, L.K. George, D.G. Blazer, and K.G. Meador. "Religion and Anxiety Disorder: An Examination and Comparison of Associations in Young, Middle-Aged, and Elderly Adults." *Journal of Anxiety Disorders.* Vol.7 (1993): 321–342.

Koenig, H.G., J.C. Hays, D.B. Larson, L.K.George, H.J. Cohen, M. McCullough, K. Meador, and D.G. Blazer. "Does Religious Attendance Prolong Survival? A Six Year Follow-Up Study of 3, 968 Older Adults." *Journal of Gerontology: Medical Sciences.* Vol.54A (1999): M370–M377.

Koenig, H.G., D.B. Larson, and S.S. Larson. "Religion and Coping with Serious Medical Illness." *Annals of Pharmacotherapy.* Vol.35.3 (2001): 352–359.

Koenig, H.G., M.E. McCullough, and D.B. Larson. *Handbook of Religion and Health.* New York: Oxford University Press, 2001.

Koopman, B.G., and R.A. Blasband. "Two Case Reports of Distant Healing: New Paradigms at Work?" *Alternate Therapy Health Medicine.* Vol.8.1 (2002): 116–120.

Krucoff, M.W., S.W. Crater, and K.L. Lee. "From Efficacy to Safety Concerns: A Step Forward or a Step Back for Clinical Research and Intercessory Prayer? The Study of Therapeutic Effects of Intercessory Prayer (STEP)." *American Heart Journal.* Vol.151.4 (2006): 762–764.

Kuhn, T.S. *The Structure of Scientific Revolutions.* Chicago: University of Chicago Press, 1970.

Lakatos, I. "Falsification and the Methodology of Scientific Research Programmes." *Criticism and Growth of Knowledge.* Eds. I. Lakatos and A. E. Musgrave. Cambridge: Cambridge University Press, 1974.

Maslow, A.H. *Toward a Psychology Being.* New York: Van Nostrand, 1968.

Mattlin, J., E. Wethington, and R. Kessler. "Situational Determinants of Coping and Coping Effectiveness." *Journal of Health and Social Behavior.* (March 1990): 103–122.

McCullough, M.E., W.T. Hoyt, D.B. Larson, H.G. Koenig, and C.E. Thoresen. "Religious Involvement and Mortality: A Meta-Analytic Review." *Health Psychology.* Vol.19 (2000): 211–222.

Miovic, M. *Initiation: Spiritual Insights on Life, Art, and Psychology.* Hyderabad: Sri Aurobindo Society, 2004.

Mueller, P.S., D.J. Plevak, and T.A. Rummans. "Religious Involvement, Spirituality, and Medicine: Implications for Clinical Practice." *Mayo Clinic Proceedings.* Vol.76.12 (2001): 1189–1191.

Pargament, K.I., H.G. Koenig, N. Tarakeshwar, and J. Hahn. "Religious Struggle as a Predictor of Mortality Among Medically Ill Elderly Patients: A Two-Year Longitudinal Study." *Archives of Internal Medicine.* Vol.161 (2001): 1881–1885.

Polanyi, M. *Personal Knowledge: Towards a Post-Critical Philosophy.* Chicago: University of Chicago Press, 1958.

Popper, K.R. *The Logic of Scientific Discovery.* New York: Basic Books, 1959.

———. *Conjectures and Refutations.* London: Routledge & Kegan Paul, 1969.

Rao, K.R. "Parapsychology: Implications for Religion and Science." *Perkins Journal.* Vol.39.2 (1986): 1–21.

———. "Psychology of Transcendence: A Study in Early Buddhistic Psychology." *Asian Contributions to Psychology.* Eds. A. C. Paranjape, D. G. F. Ho, and R. W. Riebu. New York: Praeger, 1988. 123–148.

———. "Magical Synthesis: Meditating on the Mandala of Indian Identity." Paper Presented at the National Workshop on National Integration and Multiple Identities, Indian Institute of Advanced Studies, Shimla, 15 September 2007.

Rao, K.R., and J. Palmer. "The Anomaly Called PSI: Recent Research and Criticism." *Behavioral and Brain Sciences.* Vol.10 (1987): 539–555.

Reed, P.G. "Religiousness Among Terminally Ill and Healthy Adults." *Research in Nursing and Health.* Vol.9 (1986): 35–41.

———. "Spirituality and Well Being in Terminally Ill Hospitalized Adults." *Research in Nursing and Health.* Vol.10 (1987): 335–344.

Rhine, J.B. "Telepathy and Clairvoyance Reconsidered." *Journal of Parapsychology.* Vol.9 (1945): 176–193.

———. *New World of the Mind.* New York: William Sloane Associates, 1953.

———. "Comments: Second Report on a Case of Experimenter Fraud." *Journal of Parapsychology.* Vol.39 (1975): 306–325.

Schrödinger, E.C. "Is Science a Fashion of the Times?" *Science and Society: Selected Essays.* Eds. A. Vavoulis and A.W. Colver. San Francisco: Holden Day, 1966.

Stanford, R.G. "Conceptual Frameworks of Contemporary PSI Research." *Handbook of Parapsychology.* Ed. B. B. Wolman. New York: Van Nostrand Reinhold, 1977. 823–858.

Steffen, P.R., A.L. Hinderliter, J.A. Blumenthal, and A. Sherwood. "Religious Coping, Ethnicity, and Ambulatory Blood Pressure." *Psychosomatic Medicine.* Vol.63 (2001): 523–530.

Stevenson, I. *Twenty Cases Suggestive of Reincarnation.* 2nd ed. Charlottesville, VA: University of Virginia Press, 1974.

———. "Reincarnation: Field Studies and Theoretical Issues." *Handbook of Parapsychology.* Eds. B.B. Wolman, L.A. Dale, G.R. Schmeidler, and M. Ullman. Jefferson, NC: McFarland & Company, 1977. 631–663.

Tart, C.T. "Editor's Introduction." *Body Mind Spirit: Exploring the Parapsychology of Spirituality.* Ed. C.T. Tart. Charlottesville, VA: Hampton Roads, 1997. 21–31.

Townsend, M., V. Kladder, H. Ayele, and T. Mulligan. "Systematic Review of Clinical Trials Examining the Effects of Religion on Health." *South Medical Journal.* Vol.95.12 (2002): 1429–1434.

3

DIALECTICS OF CULTURE
AND SCIENCE

R. P. Singh

The present millennium is different from earlier ones. We have scientific knowledge which is the most delicate and advanced, technology which is the most capable and sophisticated, but do we have the wisdom to make use of it such that there is still a scope for indigenous culture to leave its impression on these achievements? One of the features of human history has been that people, resources, ideas, values, and consciousness move from one place to another, and in the era of globalization these move all too fast, getting transformed by the day. But what has been the role of culture and values in such movements and transformations? Dialectics could be used to overcome/sublate/transcend the contradictions/antinomies between culture and science, between what we are and what we have.

Issues of contestation

The following questions need to be addressed and resolved dialectically:

1   Dialectics of cultural/theological lineages and modern rationalist/ scientific thought.
2   Dialectics of objectivity and subjectivity in culture and science. Science existing as an objective phenomenon is known to everyone, but science in the making is subjectively constituted, hence the notion of subjectivity, autonomy, and sovereignty need to be addressed.
3   Dialectics between faith and knowledge (culture and science).
4   Both culture and science evolved and emerged in society, hence social responsibility and accountability has to be defined.
5   Science as an institution financed either by the state or by the big bourgeoisie or by the military establishment.

Culture is the cultivation of the human mind; it is what people inherit. In 1952, A. L. Kroeber and Klyde Kluckhohn gave 164 definitions of culture. Raymond Williams in *Culture and Society* has enumerated three features of culture: culture as a way of life, culture consisting of norms and principles,

32

and finally the documentary aspects of culture such as oral/written aspects, museums, archaeology, symbols/meanings, etc. There may be negotiating aspects of culture particularly in the context of hierarchies of cultures – central/marginal, mainstream/subaltern, literate/illiterate, west/east, and so on. In view of different forms of life-world, one may take up a Wittgensteinian approach of different "language games" or Ryle's "logical geography of concepts." One may also use the hermeneutic aspect of culture where there is perpetual development of culture through cross-cultural interaction or what Gadamer says of "fusion of horizons."

It may therefore be said that human beings are culturally embedded; they grow up and live within a culturally structured world and organize their lives and social relations in terms of a culturally derived system of meaning and significance. They have either consciously adopted or uncritically accepted the culture or reflectively revised it in rare cases. As a result of European modernity, science and the scientific temper has been instrumental in molding and transforming cultural beliefs and attitudes of the people. The question that arises is whether cultural identity is getting lost in or getting replaced by scientific temper. The liberalist ideology regarded man as "given," and nature is regarded as another "given"; nature is something for man to "use," have power over, and exploit its resources for human development. The Enlightenment rationality had a dual function to perform – to validate the development of science and to vindicate values of freedom, autonomy, sovereignty, rationality, tolerance, adulthood, public and private property, etc. – and to be able to replace cultural values.

With science and its methodology, there can be an attempt to evaluate culture in terms of the content and the intent of culture, the universalistic character of culture, the hierarchical status of culture, and the pluralistic features of culture. Since different cultures represent different systems of meaning and visions of a good life, each realizes a limited range of human capacities and emotions and grasps only a part of the totality of human existence. Suppose it is said that "Everybody has freedom to live a good quality of life." Now this statement can be split into two parts – "Everybody has freedom" and "to live a good quality of life." So far as the first part is concerned, there is no contestation, but the second part is extremely contested. One may ask the question – will adhering to Christianity, Islam, Hinduism, or Buddhism without a sense of science and technology lead to a good quality of life? Will capitalism or socialism with a revolutionary sense of science and technology provide a good quality of life? Will liberalism, scientific temper, conservatism, or nationalism facilitate a good quality of life? To answer this question, first one should recognize that one culture needs other cultures to help understand itself better, expand its intellectual and moral horizon, stretch its imagination, save it from narcissism, guard it against the obvious temptation to absolutize itself, and so on. This does not mean that one cannot lead a good life within one's own culture but rather that, other things being equal, one's way of life is likely to be richer if one also enjoys access to others and that a culturally self-contained life is virtually impossible for most human beings who live in the modern globalized and interdependent world. From a pluralist perspective, no political doctrine or ideology can represent the full truth of human life. Each of them – be it liberalism, conservatism, socialism, or nationalism – is embedded in a particular culture, represents a particular vision of good life, has an application of science and technology, and is necessarily narrow and partial. Liberalism, for example, is an inspiring scientific and philosophical doctrine stressing such great values as objectivity, critical thought, human dignity, autonomy, liberty, and equality. However, they can be defined in several different ways, of which the liberal is only one and not always the most coherent.

European modernity was characterized by the desire to repudiate the past and to discard tradition, its institutions, and authority. Behind that drive was an absolute confidence in the capacity of unaided autonomous human reason to solve all puzzles and to remove the veil of mystery from reality. Reason alone can make the objective world no longer a threat to one's existence but fully subject to human control through science and technology, ignoring the past and concentrating on the present in a calculated and methodical manner.

Max Weber characterized cultural modernity as the separation of substantive reason expressed in religion and metaphysics into the three autonomous regions of science, morality, and art. Peter Berger suggested five phenomena characteristic of modernity: (1) abstraction, (2) futurity, (3) individualism, (4) liberation, and (5) secularization. Behind the separation of Max Weber's definition of "substantive reason" from the religious consciousness, and also from its basic unity, is the fundamental act of the Modern – the repudiation of the transcendent as the unifying principle and its replacement with human rationality as sovereign and as the new unifying principle of all experience and all understanding. The central and the fundamental thrust of the modern, it seems to me, is the bold and unhesitating affirmation of the autonomy of human individual and society, as not dependent on, or answerable to, any other reality. It is that affirmation that repudiates all external authority, outside human reason, whether of religions or of tradition. From that repudiation of external authority and the affirmation of human autonomy and sovereignty have come the other trappings of the Modern – e.g. modern science/technology, modern urban/industrial civilization, modern philosophy and literature, etc.

The beginnings of Modernity can be traced to that intellectual fervour that spread in Europe from the middle of the eighteenth century. The French Revolution of 1789 was a high point in the spread of this intellectual-spiritual as well as political-economic-social ferment in Western society. The process lasted from the mid- eighteenth to the mid-nineteenth century and is still spreading geographically, encompassing all cultures which adopt the urban technological-industrial system, with its capitalist mode of production and Calvinist-individualist "value-system." Culture, medicine, communication systems, educational systems, and political-economic institutions are all based on human sovereignty and autonomy. We "modern educated people" are all today, in large measure, a product of the ferment and turmoil. In India the process is pervasive but has not yet conquered all people since all have not yet been educated!

The Modern, if not identical with that process, is certainly a consequence of that intellectual-spiritual ferment which is sometimes referred to as the European Enlightenment to distinguish it from other enlightenments like that of the Buddhists, to whom perhaps the term originally belongs. Enlightenment Liberalism, with its twin children of modern science and technology and the urban-industrial society, and its outcomes, namely, the Marxist attempt to construct the ideal society and the positivist-linguistics course to capture the truth in words, is based on the affirmation of the autonomy of the human individual and his/her capacity to know, shape, and order the world. These four constitute the hallmarks of the modern.

The father of modern philosophy, René Descartes (1596–1650) was a philosopher, a mathematician, and a man of science. In philosophy and mathematics, he made supreme contributions; in physics, though creditable, they were not as good as those of some of his contemporaries. In philosophy, Descartes' outlook was profoundly influenced by the discoveries made in physics and astronomy. While it is true that he retains much of Scholasticism, such as the distinction between reason and sensibility, truth and falsehood, reason and faith, one and many, he does not accept the metaphysical-philosophical foundations laid by his predecessors but attempts to formulate a philosophical system *de novo*. This had not happened since Aristotle. Descartes is therefore rightly considered as the father of modern philosophy. In what follows, I shall discuss Descartes' two most important books, so far as its reference to science is concerned. These are the *Discourse on Method* (1637) and *Meditations* (1642). In these books, Descartes has developed his method, dualism, doctrines of *cogito*, matter and truth, innate ideas, existence of God, and so on.

Descartes begins with scepticism in regard to the senses because, "I have sometimes found that these senses played me false, and it is prudent never to trust entirely those who have once deceived us" (*Discourse* 96). In the process of scrutinizing the testimony of senses, Descartes arrives at the conclusion that it is prudent not to trust entirely anything that has once deceived us. Yet some sensory evidence, Descartes admits, are so strong that only a *madman* would doubt it, for example, that "I am here seated by the fire wearing a dressing gown." But then new doubt arises: one might be dreaming. In which case one's belief (about the fire and the dressing gown) might well be false. The upshot of the long and involved "dreaming argument" is that any statement concerning the external world may be doubted. To reinforce the dreaming argument, Descartes introduces the famous device of a malignant demon who has employed all his energies in deceiving him. He considers that the sky, earth, colours, shapes, sounds, and all external things are not more than the delusions of dreams. He also considers himself as having no hands, no eyes. To elaborate this position, Descartes says,

> Suppose therefore that all the things I see are false; I persuade myself that none of those things ever existed that my deceptive memory represents to me; I suppose I have no senses; I believe that body, figure, extension, movement and place are only fictions of my mind. What, then, shall be considered true? Perhaps only this, that there is nothing certain in the world.
>
> (*The Philosophical Writings* 19)

The *First Meditation* ends on a note of apparently universal doubt.

This basically means that material things are independent of the perceiving subject. Therein lies the philosophical foundation of physical sciences. Even Einstein, in "Maxwell's Influence on The Evolution of The Idea of Physical Reality" (1931), has said, "The belief in an external world independent of the perceiving subject is the basis of all natural sciences" (266). The philosophical foundation of physical sciences was lacking in the Italian Renaissance, but the French Renaissance, with Descartes' materialism, filled this gap.

Behind these positions, there is a deeper philosophy of European Enlightenment. There is, however, a lack of sufficiently broad, accurate, comprehensible, and usable definition of the early Enlightenment. Part of the reason for this lack is that during Enlightenment there have been complex and quite often contradictory views on such issues as democracy, modernity, secularism, religion, and scientific knowledge, etc. It is very difficult to provide one definition as the definition of the Enlightenment, which fits all the men usually assumed to belong to it. Generally among the Enlightenment thinkers we have Voltaire, Rousseau, Hume, Condorcet, and others. Without going into the details of their specific philosophical systems, their mutual agreements and disagreements, let us focus on freedom, rationality, autonomy, sovereignty, toleration, and adulthood as the key concepts of Enlightenment. Notwithstanding the mutual differences between one philosopher and another of the Enlightenment age, all of them have a fundamental preoccupation with freedom. It was Kant, one of its earliest prophets, who asked that question and answered it in his article in the *Berlinischer Monatsschrift*, December 1783 issue, entitled *Beanwortung der Frage: Was ist Aufklaerung? Or* "Answer to the Question: What is the Enlightenment"? His answer is: "*Aufklaerung ist der Ausgang des Menschen aus seiner Selbst-verschuldeten Unmuendigkeit.*" His full answer in English may be translated as:

> Enlightenment is the coming out of man from his self-imposed immaturity. Immaturity is the incapacity to serve one's own understanding without direction (*Leitung*) from another. This immaturity

is self-imposed; Reason itself languishes, not because it lacks understanding; what it lacks is resolution and courage; it is unwilling to
serve itself (*Sapere Aude! Hebe Mut*). Take courage to serve your
own understanding! This is therefore the Motto *(Walspruch)* of the
Enlightenment.

(translation mine 9)

It is in this rather general framework of the Enlightenment rationality that
the concept of humanity has evolved, and it gets its elaborations in the categorical imperatives.

Until Enlightenment, the integrating intellectual principle was the belief
in God. It was in theology that all human problems in experience were integrated. The Enlightenment rejected that integrating principle – the religion
as the matrix of thought process. In that place the Enlightenment put human
reason which could integrate everything. This was the basic change which
European Enlightenment brought. Once you subscribe to reason, you find
that the integrating principle does not fully work. So one can divide "experience" into three compartments – science, ethics, and art. In the Enlightenment, technically it is human reason that reconciles the three. But that
integration is very flimsy and does not have adequate foundation. Immanuel
Kant particularly tried to distinguish between three kinds of reason – pure
reason, practical reason, and judgment. In the first, you know the things
(phenomena); in the other, you know how to act; in the third, you have to
discern what is good. By making this separation, he held on to the "idea
of reason," which was already divided into three compartments. European
Enlightenment had this problem that "reason" as such is not able to fulfill
the task of integrating everything. But the Enlightenment was able to assert
on the "autonomy" and "adulthood" (maturity). According to evolutionary
biology, humanity has been developing into three phases; one is the religious
stage, and the second stage is metaphysics. These two stages are the stages of
"immaturity" of humanity. Humanity becomes "mature" when its knowledge becomes "scientific," which is the third stage. Science is the mature
form of humanity dealing with reality. Both religion and metaphysics belong
to the "childhood" of humanity. Maturity means repudiating religion and
metaphysics. The positive thing is that it affirms humanity. The attempt to
get rid of "self-imposed immaturity" is both self-critique and self-reflection
with the aim to attain emancipation. Emancipatory self-reflection is dependent on giving a rational reconstruction of the universal conditions for reason. To use the Kantian analogy, only when we understand the possibility,
validity, and limit of theoretical knowledge and the categorical imperatives
does it become intelligible to specify what must be done to attain autonomy
and emancipation. This immaturity is self-imposed, because reason itself
languishes not in the lack of understanding but only in the lack of resolve
and courage to serve oneself without direction from another.

37

In other words, the Enlightenment develops reason to the extent that it becomes autonomous and gets rid of the restraints of tradition and authority. The way to enlightenment, Kant emphasizes, is not to seek a mentor or an authority in thinking, in willing, and in feeling. Kant has placed freedom and maturity *(Muendigkeit)* at the center of enlightenment and contrasted it to tutelage. In an uncharacteristic manner, Kant says, "when the question is asked: do we live in an enlightened epoch *(Zeitalter der Aufklaerung)* then the answer is: No, but rather in an epoch of Enlightenment" *(Zeitalter der Aufklaerung* 96). Reason, the supreme faculty, has an emancipatory goal. But this is contrasted by the *a priorism* of the faculty of understanding, which can vindicate only a limited theory.

Modern philosophical reflection, especially after Kant, is preoccupied with "man": as Gutting puts it, "before the end of the eighteenth century, Man did not exist and . . . he will disappear with the (apparently imminent) collapse of the modern episteme" (198). To develop a modernist notion of man, I would like to turn to Richard Rorty's consideration of the dispute between Cardinal Bellarmine and Galileo. For, as Rorty explains, "Much of the seventeenth century's notion of what it was to be a "philosopher" and much of the Enlightenment's notion of what it was to be "rational" turns on Galileo's being absolutely right and the Church absolutely wrong" (328). In modernism, Bellarmine's appeal to Biblical scriptures to limit the scope of Copernican theory is seen as illegitimate in so far as it imposes a religious dogma on a scientific hypothesis. In other words, it imposes non-scientific values on purely scientific concerns and thus fails to understand the distinguishing marks of rational knowledge as opposed to faith. The conflict between science and religion is itself a historical product of modernism. This conflict did not exist before Galileo's defenders used it to refute Bellarmine. We support Galileo because we are his heirs.

> We are the heirs of three hundred years of rhetoric about the importance of distinguishing sharply between science and religion. . . . But to proclaim our loyalty to these distinctions is not to say that there are 'objective' and 'rational' arguments for adopting them. Galileo, so to speak, won the argument and we all stand on the common ground of the grid of relevance and irrelevance which modern philosophy developed as a consequence of this victory.
>
> (Rorty 330–1)

Galileo's reply was his *Letter to the Grand Duchess Christina*, in which he argued for the strict separation of theological and scientific issues on the ground that science and religion require different enterprise so that the truth of science should not conflict with the truth of religion.

Kant, taking up these issues in further detail, tried to limit knowledge in order to leave a room for faith. Kant defines scientific knowledge as synthetic *a priori*. As synthetic, it amplifies the concept of subject and is more than a mere tautology. As *a priori*, it expresses universality and necessity. What we need in science, according to Kant, is such ampliative knowledge with the characteristics of universality and necessity. But scientific knowledge, Kant warns, has a limit. It is limited to the phenomenon, and there is a noumenal world where science cannot penetrate. Noumenon is the sphere of faith and morality. Man is a member of both the spheres – science and morality. This is explained in the Third Antinomy of the *Critique of Pure Reason*.

> In the thesis, nature is taken as the unity of all objects of possible experience and is determined as a whole by laws. These are the laws of causality, which decisively explain a cause. Every event in the world of Nature is caused by a preceding event, and that in turn is preceded by another. There must be a first cause, which is not further caused by anything else. This, Kant says, is free causality. Kant here brings out his notion of will in order to become practical and transforms itself into empirical consequences.
>
> (Singh 145)

The solution of the Third Antinomy discovers the possibility of causality through freedom in harmony with the Universal Law of natural necessity. Man, in this exposition, must be viewed as "On the one hand phenomenon, and on the other hand, in respect of certain faculties the action of which cannot be ascribed to the receptivity of sensibility, a purely intelligible object" (*Critique* 472). St. H. Watson explains,

> In this respect the constitution of man is the coefficient of the two types of causality. Man is neither phenomenally free, nor noumenally conditioned. As an appearance he stands with all other appearances and therefore subject to the laws of natural necessity. But, as noumenon, he stands apart from nature. In a sense both types of causes are affirmed as possible in one event. The noumenal is outside the series of appearances and yet possessed by a causal agent that also is an appearance. Therefore man, as agent, may have both an intelligible character and an empirical character.
>
> (Watson 168)

Being intelligible, man is an ethical being. Being empirical, man is a scientist. Kant prescribes no fewer than five formulas of the categorical imperatives, and it would exceed the limits of our present study to make a detailed examination of his analysis of each of these formulas.

To conclude, it can be said that these are some of the most general features of scientific rationality, being both intellectual and perspectival. These features are claimed to be both instrumental and emancipatory at the same time. For Indian minds puzzled about modernity and the science and scientific temper of Europe, I humbly recommend a change of perspective. Let us leave Descartes, Kant and Hegel, Marx and Freud for a while, get out of the Enlightenment frame of mind and go to the Upanishads. There is no other way of detoxifying ourselves from the fumes of Enlightenment rationality. Because the Western way is not the only way of thinking and experiencing, let us as Indians immerse ourselves in our own rich heritage, especially before its breaking up into Buddhist, Jaina, and Hindu – for example, the Samkhya Yoga heritage common to all three traditions – the great philosophical perspective that undergirds all Upanisadic, Vedic, Buddhist, or Jaina thought and experience. Keep your painfully acquired critical rationality, but do not get tyrannized by it. Stay critical, but do not reject out of hand what seems strange at first. Expose yourself without hesitation to a system of thought and experience which has endured for millennia.

Works cited

Berger, Peter. *Facing Up to Modernity*. New York: Basic Books, 1977.

Descartes, Rene. *Discourse on Method and the Meditations*. Trans. E. Sutcliffe. New York: Penguin, 1979.

———. *The Philosophical Writings of Descartes*. Trans. John Cottingham, Robert Stoothoff, and Dugald Murdoch. Vol. II. New York: Cambridge University Press, 1984.

Einstein, Albert. *Ideas and Opinions*. New Delhi: Rupa & Co., 1989.

Gutting, Gary. *Michel Foucault's Archaeology of Scientific Reason*. Cambridge: Cambridge University Press, 1990.

Kant, Immanuel. *Critique of Pure Reason*. Trans. N.K. Smith. London: Macmillan Reference, 1973. A547, B575.

———. *Was ist Aufklaerung: Thesen und Definitionen*. Reclam: Stuttgart, 1986.

Kroeber, A.L., and Klyde Kluchohn. *Culture: A Critical Review of Concepts and Definitions*. Cambridge, MA: Peabody Museum of American Archaeology, 1952.

Rorty, Richard. *Philosophy and the Mirror of Nature*. Princeton: Princeton University Press, 1979.

Singh, R.P. *Dialectic of Reason: A Comparative Study of Kant and Hegel*. New Delhi: Intellectual Publishing House, 1995.

Watson, St. H. "Kant on Autonomy, the Ends of Humanity and the Possibility of Morality." *Kant-Studien*. Vol.77. Jahrgang Heft 2 (1986).

4

CULTURE AND SCIENCE IN TWENTY-FIRST CENTURY INDIA

V. V. Raman

This chapter will discuss the aesthetic and ethical dimensions of culture; it will also discuss worldviews, especially their two important components – the physical phenomenon and the experienced conscious. It will explore these largely in the Indian context. The chapter will also critique the postmodernist interpretation of modern science and argue that India's contribution has been, and will be, significant in the second component of worldviews, that is in the experienced conscious.

Every biological species on our planet is unique in its own way. That all human beings belong to the same species is a scientific truth. In religious-cultural terms, we are all children of the same divine principle; we all belong to the same human family.

It is a fact of the human condition that, unlike most other species, we form groups and subgroups which are sometimes mutually cooperative and sometimes unpleasantly combative with others. Humanity is a mosaic of countless pieces on the basis of three major factors: ethnicity, religion, and language. These powerful elements both unite and divide us all as cultural beings. They are the primary building bricks of cultures. In our own times, because of their intermingling in non-homogeneous nations, cultures have become disfigured by politics, hegemonies, rivalries, and chauvinistic breast-beating.

Race, like caste, is an outmoded concept, even a hurtful one, because it gave rise to some pernicious practices. We rather talk about *ethnic identity* these days. Call it what you will, human beings do have superficial dissimilarities, and we still refer to people as white, black, brown, or yellow. What matters is not how we classify human beings on the basis of the colour of their skin, the religion they profess, or the language they speak but how we regard and treat fellow humans.

Our cultural experience is strengthened by our emotional attachment and allegiance to traditions, festivals, and pride in ancestry. We may respect and appreciate other cultures, but one's culture has no meaning if one's heart does not resonate with it. Culture demands both understanding and affiliation. Thus, Bengali citizens in Canada are Bengalis first, then Canadians;

British Nigerians are Africans first, then Britons; Arabs in France are Muslims first, then Frenchmen. Language, religion, and race are powerful indeed, and they triumph over the passport we carry. Culture finds expression in a multitude of ways, but its three important dimensions are aesthetics, ethics, and worldviews.

Aesthetic dimension of Indian culture

Matthew Arnold said that culture is "to know the best that has been said and thought in the world." I would go further and say that it is the best and the noblest that has also been created and given form and sound in humanity's heritage.

The aesthetic dimension of culture is its most enjoyable aspect. It includes music and dance, play and poetry, cooking and cuisine, and in our own times, movies too. Historically, in all cultures these were often linked to the religions that inspired and enriched the people of a land. But now it does not necessarily have to be so.

India's culture is uncommonly rich and multifaceted and also extraordinarily diverse. In few other nations of the modern world does one find such an amazing range of languages and dialects, images and symbols, as well as costumes, customs, and culinary variety. The people of India, in whose veins course the blood of every race and religion, represent practically every genetic group. This may explain their creativity, complexity, and capacity for diverging discourses.

The aesthetic expressions of Indian culture include magnificent sculptures in the countless temples that grace India's landscape, from miniature *vigrahas* in simple shrines to mammoth rock sculptures immortalizing episodes from our epics. Not just the majesty of the dancing divine in Chidambaram or the fury of the Goddess of Kalighat but every *murti* in a Hindu temple may be seen, beyond its spiritual essence, as also a meticulous work of art. But there is something incomplete in all this: the names and dates of practically all the great sculptors of classical India were seldom inscribed in the archives and so have faded away with the memories of distant generations. We don't have a Michelangelo or Bernini to recall when we stroll through Mahabalipuram, stand in admiration at the fantastic carvings at Khajurao, or marvel at the remains of the Buddhas of Bamiyan. Then there is all the magnificence of Indian music, from the simple melodies of village folk songs to the weeklong music festivals, that draws thousands here in India and beyond our shores. The Indic musical tradition goes back, as we all know, to the serene hymns of the Samaveda and have emerged in countless complex *ragas* through an impressive range of instruments from *bansuri* and *mrdangam* to *veena*, *sitar*, and much more. Islamic contributions to Indian music, both in composition and in execution, are considerable. So are its majestic monuments in architecture. In more recent times, enrichments have also come from the cultural treasure chests of the Western world. Indian culture is rich because it has always been open, welcoming, and transforming.

Then there are the grand epics which have breathed life into the Indic civilization and beyond and also inspired vast outpourings of poetry, music, and dance in India. Add to these the stories and plays that adorn every language spoken in the country, and the ones in English too, and we have an idea of the breadth and range of Indian culture.

As to the religious splendor that lights up the Indian spirit, here arose, as Swami Vivekananda once thundered, the doctrine of the immortality of the soul, the existence of a super-immanent God in Nature and in Man, and here the highest ideals of religion and philosophy attained their culminating point. In our own times, though the vast majority of Indians are Hindus by birth and cultural connection, Hinduism itself is a mighty complex system with a thousand different shades and sects that range from pure materialism and stark hedonism to adherence to ancient convictions. But here may also be found some of the most sublime instances of spiritual awakening in the world.

The citizens of this great nation are affiliated to practically every faith and creed: not just Jains and Buddhists and Sikhs of Indic vintage but also Parsees, Muslims, Christians, and Jews, and countless others who, either overtly or secretly, consider themselves hard-core humanists, skeptics, and atheists. By and large, the people of this culturally blessed land coexist in happy harmony and share their respective richness with others. Not just India's intrinsic character but her glory and greatness lie in this unsurpassed diversity of faiths and abundant variety of creeds much more than in her material resources, which are considerable. The exemplary accommodation of ideas and worship modes that has generally existed in this land may be seen as reverberations of the wisdom of the Vedic sage-poet who declared pithily that there is but one truth and that it is articulated in multiple ways. This is a precious gift, this openness, and we may hope that we will never forget or abandon it. Indeed, we should strive to preserve it with all our love and commitment. India should offer this deep insight to the world troubled and tarnished by religious fanaticism, intolerance, and hatred of the other.

Fusions and confusions

The distinctiveness and local evolution of cultures in different countries gave rise to immense richness for the human family as a whole. Kalidasa and Kamban, Dante and Cervantes, Shakespeare and Goethe, and such others gave special luster to their respective languages, as did Thyagaraja and Tagore, Bach and Beethoven to the music of their traditions. There is a living spirit in every culture, a uniqueness with deep roots, and that uniqueness has local fragrance like the flora and fauna of a geographical zone.

Up until recently, the ancient roots stayed separate and sturdy, and the emergent trees grew taller and vigorous too, shooting out branches along different directions. Moreover, all the branches of each cultural tree drew nourishment from the same indigenous roots. Whether in art or poetry, music or dance, different sects and schools arose in different regions, but in each instance, even when there were interactions and influences, the core stayed safe and secure.

In the global context in which we live, it is both inevitable and to an extent appropriate that cultures mingle and mildly modify one another. Already in ancient times such influences existed. In the eighteenth century, Goethe and Schiller and Schopenhauer in Germany, to mention just three, were moved by Kalidasa and the Upanishads, and in the following century the likes of Ralph Waldo Emerson and David Thoreau were touched by the Bhagavad Gita. Conversely, many nineteenth-century Indian writers were enriched by European novelists. Nalappat Narayana Menon's translation of *Les Misérables* into Malayalam, like numerous other translations of English and French literary masterpieces into various Indian languages, changed the course of literary history in India.

However, all that was different from the so-called fusion experiments that are going on today. There is surely charm in Yehudin Menuhin and Ravi Shankar fusing to create a new kind of music and initiating a synthesis thereby, as there is colourful entertainment when Bollywood imitates Hollywood. But sometimes things can be carried too far. Indeed Western, and especially American, cultural sweep strikes many thoughtful people as more detrimental than desirable.

Non-Western peoples tend to think that this is a problem they alone face, but in fact people in France and Italy, in Finland and Russia, and in other Western countries also fear that their local cultures are being diluted, tainted, and even obliterated by an all-devouring Americanism whose crass manifestations include McDonald's and Kentucky Fried Chicken, Coca Cola and Wal-Mart. These and MTV are either embraced in their totality or zealously aped in many countries, often marginalizing regional varieties of food and drink, music, art, and dance.

Sometimes these result in clumsy caricatures, as when *aloo-roti* is transformed to *aloo-pizza* and restaurants offer chicken *masala dosa* or spaghetti *semya*. It will be a sad day if Kali puja is turned into a Woodstock festival of hippies, and *bhajans* are sung with guitars and the electronic keyboard. The consequences can be both comical and catastrophic. Such metamorphoses will not only discolour but also distort cultures that have withstood the shocks of centuries while evolving themselves within their confines, stirred by indigenous creativity, and not disfigured altogether by hegemonic thunderbolts from extraneous sources. Blind transplantation of alien cultural modes will most likely replace in totality what is intrinsic to a culture.

Ethical dimensions of culture

The second important aspect of culture is the ethical dimension. Traditionally, in most civilizations the ethical framework arose by the religions of the people. In the Ten Commandments, the Sermon on the Mount, the Bhagavad Gita, and the Sharia, for example, we see the source of traditional morality. Most of the injunctions formulated here relate to our conduct towards fellow humans, to self-restraint, and to our attitudes towards God. They are not abstract ideas of a universal nature, except in some prayers like *loka samasthâ sukhino bavanthu.*

In this regard, there have been important awakenings of a collective human consciousness as a result of what is generally called the Enlightenment in the West. The term has acquired a pejorative connotation in postmodernist circles. Be that as it may, emerging from modern scientific perspectives are values like a skeptical attitude towards human knowledge, rejection of the infallibility of texts and authorities, non-acceptance of absolute truths, respect for carefully collected facts and figures, suspicion of simplistic explanations, and profound respect for observational evidence that is supported by reason and logic. These, combined with dedication to an unfettered quest for understanding every facet of the experienced world and of human history, recorded and submerged, tend to rid the human mind of the natural fear of the unknown, of obscurantist mumble-jumble and time-honoured magic-mongering, not to mention the elimination of some awful social and political evils, such as dictatorship, slavery, gender oppression, exploitation of the underclass, and the like.

The ideals of universal justice and equality, the striving for the improvement of the human condition at large, the vision of social and political problems in global rather than in parochial terms: all these go hand in hand and are in perfect harmony with what may be called an enlightened outlook. The universality implicit in the scientific search and the recognition of the historical roots of culture and religion are preconditions for attitudes that transcend sub-cultural barriers. The concepts of a chosen people, of a personal monolingual God who made a specific geographical location his hallowed spot and a select set of languages sacred, of superior and inferior castes, of a master race: these and similar constraining convictions lose their appeal to the enlightened mind.

Culture and worldview: science

The third important dimension of culture pertains to our worldview. And worldview has two distinct components: the first pertains to our picture of the physical world and the second to our understanding of the roots of the human mind and of human consciousness.

The history of human civilization is marked by several revolutions, some slow and some abrupt, some dramatic and some subtle, some of local significance and some of global impact. Among the most important of these are the agricultural revolution, which introduced sowing, harvesting, and storage of crops; the cultural revolution from which emerged abstract thoughts and ethical frameworks, as well as philosophies and religious systems; the scientific revolution, which changed the coordinates of our planet from cosmic center to an insignificant niche in an immeasurably vast expanse; and the industrial-technological revolution, which harnessed matter and energy on the basis of an understanding of the workings of the physical world.

It would be a misreading, not to say distortion, of history to say that in earlier times there was neither science nor technology. From the unrecorded dawn of consciousness, when the human mind wondered and human hands turned a stone or a stick this way and that to feel and fathom what it was, science has been there in every community and culture. In periods now long past, scientific creativity and discovery flourished in India and China, in Egypt and Mesopotamia, in Greece and elsewhere. Devices have been contrived to lessen muscular effort and facilitate human manipulation of the world since time immemorial. Wonderment and curiosity about the surroundings and eagerness to diminish sweat and work are inherent to the human spirit.

However, what occurred in the sixteenth and seventeenth centuries was in every sense a new approach to unscrambling the complexities of the phenomenal world. The new science that emerged relied much more on careful observations, on the use of specifically constructed instruments such as lenses, telescopes, and microscopes, and also on an increasing use of quantitative methods and mathematical analysis.

The scientific revolution of the sixteenth century was significant not so much in the discarding of geocentrism, though this was one of its earliest steps; not so much in the discovery of elliptical planetary orbits, though this opened our visions to hitherto hidden aspects of the universe; not even so much in the formulation of the laws of motion, though these led to a deeper understanding of the physical world. It was significant because it initiated a universality, which has transformed the very nature of the scientific enterprise.

Since the emergence of modern science, the enormous range of scientific efforts in different countries, and then in different continents, have come to be subsumed under a single umbrella made up of an abstract international body of scientific practice and culture. The various nations of the world have their own research laboratories and publications, and yet the works carried out and published in these geographically separated places are interwoven into a web held firm by invisible bonds that know no borders, that feel no cultural differences. Meter and kilogram in any national bureau of standards are precisely the same no matter what religion or form of government may be followed in the country.

Science certainly has its local interests, narrow nationalism, and petty fights over priorities too. After all, it is only a human enterprise. There are rivalries and races in the pursuit of knowledge and competition in discoveries. There is national pride when an international prize is announced. And yet the technical work of scientists is blind to nationalities, they overlap and mingle like sounds from different instruments in an orchestra to create and constitute the grand symphony that science is. The true strength and stature of modern science lies in its universality. Science is no longer bits of insights here and there, nor imaginative speculations undertaken by curious minds in particular cultures. It surely is not parochial ethnic interpretations of natural phenomena, or narratives from sacred books. Science is a collective quest, a restless drive to eradicate every misunderstanding in the interpretation of every occurrence from the micro to the macrocosm, to unravel every mystery and dispel every doubt and darkness from the inquiring mind.

In no other context: not in art, music, sports, much less in politics, do men and women of all races, languages, and religions hold hands as comrades in a common pursuit. This speaks as much for the glory of the science as an enterprise as all its technological triumph does. Whether one is from the East or the West, from the North or the South, of India or of the globe at large, modern science is not Western any more than zero is Indian or gunpowder is Chinese, except in the accident of their geographical origin. For, for better or for worse, the scientific revolution has merged diverse streams of search into a single surging river, as it were.

What characterizes modern times is not only transnational science but also the ubiquity of modern technology. There is no member state of the United Nations organization where science is not taught or planes don't land. In spite of all our national differences and cultural diversity, no matter what language we speak and what creeds we subscribe to, the one common thread that connects the minds of men and women in today's world is international science. So, too, the commonalities in the towns and cities of the modern world are electric lights and communication systems, automobiles and computers, all unfortunately products of modern science.

Science and technology hold the sway in our world. If we look around any spot on earth that has found its way into the mainstream of human history, we cannot escape the presence of wheels and wires, of gadgets and generators, of vaccines and pills. The material impacts of science, the magic and madness of machines, are omnipresent and inevitable. Technology is certainly here to stay, and its influences are likely to grow even more in times to come. Technology can flourish without science. For instance, the copying of prescriptions, the construction of factories, and the training of technicians that run these factories can be done with very little science.

Postmodernist challenges to the universality of science

The postmodernist movement which emerged in the West in the last century has had many positive impacts on human values and perspectives. It recognized the intrinsic worth of all cultures, insisted on the dignity and self-respect of all civilizations, and rejected hegemonic attitudes in any field of human endeavor. Cultures are no longer classified as primitive and advanced; no religion is intrinsically superior or inferior to another. These are among the many contributions of the postmodernist movement.

In this framework, some postmodernist thinkers began to argue that no judgment on an issue is necessarily better or worse, that no answer to a complex question is right or wrong, and that science has no absolute validity, that it is essentially a cultural construct devoid of the objectivity it claims. This thesis is not without some merit, and it has been enthusiastically embraced by many in the non-West.

Those who rebel against this globalization of scientific worldviews may do so at the philosophical level. They may decry the claims of objectivity of modern science, but they cannot communicate their views to the world without using the devices and instruments that have been constructed only by the application of modern scientific knowledge. Cultures and countries that reject the internationalism of modern science will do so at their own peril. They will be kept in a permanent state of disadvantage vis-á-vis more materially advanced and scientifically awakened nations. Indeed, this narrow vision of science, if seriously adopted beyond the academic ivory tower in emerging nations, could affect the non-West adversely in immeasurable ways.

It is fortunate that right from the start, already from the last decades of the nineteenth century, wise leaders and enlightened thinkers in modern India embraced modern science, with the result that today thousands of Indians are making significant contributions to the world of science, both from within the universities and laboratories of India, and from beyond, in practically every scientifically advanced country. They are also contributing to science education in many developing countries of the world.

It is important to understand that modern science was a revolutionary change in humanity's approach to the understanding of the world; its framework has little to do with technological gadgetry. It is the modern scientific attitude that makes archeology, dispassionate history, and cultural anthropology possible. It is the scholarship and curiosity emerging from modern science which deciphered hieroglyphics and cuneiform tablets, brought to light the glories of ancient Egypt and Sumeria, unearthed Mohenjodaro, and more.

Mind and consciousness

In the other dimension of worldviews, namely the exploration and unscrambling of the nature of the human mind and the enigma of human consciousness, Indian visions, both classical and modern, could have a significant role to play.

The vast corpus of classical writings is often regarded as another interesting body of speculative thought about the world and human consciousness. In this view, Indian philosophy is rich in the variety of problems it explores, impressive in its scope and range. It may strike us as creatively imaginative in its analogies and hypotheses, tantalizingly appealing in its picturesque worldviews. Few who have even scratched the surface of the grand visions of Indian philosophy can deny that the thinkers who originated them were mighty intellects.

From another perspective, however, even with mutually opposing positions as to the identity or distinctiveness of the *jivatman* and the *paramatman*, and other mutually irreconcilable metaphysical assertions, Indic reflections have at their core certain profound insights into the ultimate nature of ultimate reality and the human experience. Their basic theses do not simply subtend speculative systems any more than Maxwell's theory of electromagnetism is mere mathematics. Rather, Indian visions are telling us something that is not only meaningful but revelatory about the nature of consciousness and the cosmos. They are not building a system of philosophy so much as unveiling a not-so-apparent dimension of Reality with a capital R. Their assertions were not just doodles on the mental plane: they arose rather from experiential certitudes resulting from sustained investigations of the subtlest centers of the inscrutable Self. Their words and wisdom are to be taken not as grand imaginative poetry but as findings and discoveries about the physical universe, exactly as twentieth-century science, after persistent probes into the heart of matter and energy, after countless hours of search and reflection, has erected its own views of fundamental reality.

If this were so, if spiritual probing via yogic techniques do lead to insights about aspects of the nature of physical reality, while scientific peelings of the layers of matter via experimental ingenuities and mathematical formalisms lead also to the deep-down details of that same reality, then one would expect the two lines of quest to merge, somewhat as travelers by jet planes and by ocean liners, starting from the same point, could ultimately meet at the same destination.

This, in the view of some, is precisely what is happening in our own times. For it turns out that the philosophical quagmire into which quantum physics has been sliding during the past few decades turns topsy-turvy our common-sense pictures of a solid substantial world of cause and law, of rigid particles and conserved quantities, of smooth flowing time and three-dimensional spaces. As we delve deeper into the remote recesses of atoms and nuclei, funny things begin to happen: mathematical clouds of probability take over, electrons seem to know, photons are entangled and information seems to get transmitted instantaneously, everything shows signs of being interconnected, and a good many more strange things are taking place in the microcosm. In the depths of black holes and in the singularities of quarks, space and time and physical laws themselves get warped and dissolved.

Now we begin to wonder if those *rishis* of ancient India had not after all tumbled upon some profound truths about the perceived world which, because of their very nature, could not be adequately expressed even in Sanskrit. They were perhaps quite right in insisting that in the stark denuded aspect, stripped of mute matter and measuring mind, there is a level of reality that only pure consciousness can experience, and pure consciousness can only experience, not convey. Could it be that now at long last, after countless tortuous turns of reason and experimentation, of mathematics and microscopes, science is slowly beginning to get a glimmer of what the sages were speaking about?

Quite a few thoughtful people are persuaded by this possibility. That is why some physicists and philosophers of the quantum world, commentators, and speculative thinkers are drawn more and more towards ancient insights. It would seem, as Alex Comfort suggested in his *Reality and Empathy*, that there is much to be gained if the yogic quest on the one hand, stripped of its mumble-jumble, and no-nonsense empirical science on the other, stripped of its rationalistic straight-jacket and model-building mind-set about what can and what cannot be, combine forces in unscrewing the deeper mysteries of the world of experience.

India is fortunate in its ancient cultural heritage. The aesthetic components of that culture are colourful and growing. They are being constantly enriched by interactions with other cultures of the world. It is important that in this process, Indian culture does not dissolve into a larger amorphous global culture rooted mainly in the West and dominant by virtue of its relative political and economic strength. The ethical dimensions of Indian culture are a mixture of ennobling principles as well as anachronistic practices. Here it is important, indeed necessary, for spiritual and lay leaders to direct the people along intrinsically enlightened modes informed by enlightened values and discard aspects that are both unconscionable and unacceptable. These include the treatment of certain classes of society and also the place given to women in actual practice, as against enunciated theory. In this context, India has been doing well in joining the evolving global standards of vice and virtue that go beyond parochial and theocratic confines: freedom for all, human rights, democratic form of government, and the like. Finally, at the worldview level, on the one hand we should celebrate transrational beliefs about God and afterlife and also investigate analytically and with scholarly depth what aspects of ancient visions lead us to deeper understandings of ultimate and mind-related issues. At the same time, it is important to distinguish these from naturalistic explanations of phenomena in the physical world offered by the meticulous methodology of modern science and modern mathematics. These transcend race and religion, geography and nationality, and to which a growing number of people of Indic origin have been making significant contributions.

5

MODERN SCIENCE IN INDIA AND THE EMERGENCE OF A NEW WORLDVIEW

Challenges and opportunities

Job Kozhamthadam

I begin with the understanding of culture as "the sum total of all the devices and methods that a society has at its disposal to control, direct, and modify the material conditions of its existence." Never stagnant and ever living, culture is the dynamic mechanism that enables a society to respond responsibly to the challenges arising from the various developments of the day in order to creatively and constructively transform them into opportunities to grow, develop, and mature.

Thanks to the unprecedented developments in science and technology in recent times, world culture in general, and Indian culture in particular, is in the throes of a paradigm shift challenging the usefulness of several of the existing principles and values, reaffirming some of the old ones, and inviting to induct or appropriate new ones. All this new turn of events is paving the way for the emergence of a new world order, and India is a major player in this global phenomenon. I argue that a creative and constructive dialogue between modern science and spirituality can be of immense assistance in this crucial process, particularly in ensuring that the process proceeds along a trajectory that ensures the wholesome development of humanity and of the universe around. Furthermore, I believe that India is in a privileged position to serve as a catalyst in this process since she holds in high esteem both modern science and spirituality.

In the past, science was very much looked upon as a provider of important but dispensable amenities and comforts of life. However, today it is being recognized that science has and does play a far deeper and wider role in our world and society, thanks mainly to recent developments in science and technology, particularly in the field of the biological sciences. It can be said that today modern science is in a position to have an important say in determining not only what we have and want to have but also what we are and want to become. Not only is modern science an integral

51

constituent of our culture, it also is slowly but surely reshaping our culture, paving the way for the emergence of a new world order.

India has always been the home of an ancient and rich culture, noted for its wide variety and deep spirituality. Today modern science is daring to reshape India's deep-rooted values and time-tested customs. This chapter is a critical study of certain developments in modern science and technology which exert a profound influence on contemporary world culture in general and Indian culture in particular. After discussing briefly some of these important developments, I will point out that these changes have come to stay, and there seems to be no turning back. A new order is taking definite form in the world scenario, and a new culture is emerging in India, with all its "well effects" and ill effects, offering serious and almost unavoidable challenges. It is for us to transform these challenges into genuine opportunities in order to usher in a richer and nobler world. I will further argue that a creative and constructive interaction between modern science and spirituality can be of considerable help in this laudable venture by maximizing the "well effects" and minimizing the ill effects. Furthermore, India is in a privileged position to contribute substantially towards this dialogue since here both science and spirituality have always been taken very seriously.

Some recent developments in science and their implications

A number of developments initiated or catalyzed by developments in science have played a pivotal role in bringing about the new world order. I discuss below some of the more important ones.

Shift from the age of discovery to the age of mastery

Thinkers are pointing out that today science is moving from an age of discovery of nature to an age of mastery over nature. In the past, scientists were pleased and felt fulfilled when they succeeded in revealing the secret laws of nature. Johannes Kepler, Isaac Newton, Albert Einstein, etc. were acclaimed as outstanding scientists because of their discovery of important laws of nature. Today scientists seem to be far more ambitious: not only do they want to make new discoveries, they also want to have mastery over the operations of nature and to have a hand in determining the destiny of nature. A good illustration for this new focus is the Genetic Revolution that is unfolding before us, conspicuously from the second half of the twentieth century. Certainly this revolution involves the discovery of new laws. However, today we see that these laws are being skillfully transformed into tools of mastery to reshape the nature and destiny of humans and the world around.

Paradigm shift in the role of science

Closely linked to the shift of focus is the transformation taking place in the role of science and the impact science has on society. As mentioned already, in the past the role of science was confined very much to providing certain amenities and comforts in life. Computers, TV sets, mobile telephones, etc. were some of the outstanding gifts of modern science to humanity. These contributions certainly helped to enhance the quality of life, at least for those who could afford them. But they remained optional: one could remain neutral towards them without being affected in any significant way. They touched humans from the outside; they concerned the world around humans. But today science is in a position to go deeper, to human persons themselves. For instance, the main actors of the Genetic Revolution – genetic engineering, the Human Genome Project (HGP), cloning, nanotechnology, etc. – touch human beings themselves in a significant way. We may say that so far science has dealt with the conditions of life, but today, thanks to developments like the Genetic Revolution, science is in a position to deal with life itself. So far, science has focused on what humans have and want to have, but in the Genetic Revolution it focuses on what humans are and can be.

The narrowing of the geographical boundaries and the formation of the global village

The term "global village" is closely associated with Marshall McLuhan, who popularized it way back in the 1960s in his well-known books like *The Gutenberg Galaxy: The Making of Typographic Man* in 1962 and *Understanding Media* in 1964. Although in the past it was used as a metaphor to describe the amazing speed with which electronics and the Internet could interlink different parts of the world, today it is becoming more and more literally true. Our globe has become a village in the true sense of the word – whatever happens in any part of the world, however remote and inaccessible, is almost instantly made accessible to anyone, anywhere. This revolution cannot be passed over as a mere advance in the ease and efficiency of communication; it has very serious implications and far-reaching consequences – social, cultural, political, economic, religious, etc.

Tension between global and local interests: emergence of regional nationalism and power centres

The present world scenario is full of inconsistencies and contradictions, apparent as well as real, giving rise to tensions and uncertainties. One such instance is the tension between global and local interests. Never in the history of our globe has there been so much progress made in the process of globalization and widening of interests as well as concerns. At the same time one sees local groups with their own specific agenda clamouring for world attention and support. The aborigines (*adivasis*, first settlers), the Dalits, and other disadvantaged groups in various parts of the world are raising their voices for a sympathetic and decisive hearing. There is no doubt that the process of globalization is here to stay; there is no turning back. The challenge facing our world today is to discern and identify appropriate ways of responding to these local needs without jeopardizing the process of globalization.

Greater realization of the need and relevance of collaboration and teamwork

One area in which the positive impact of the globalizing tendency is noticeably seen is the growing awareness regarding the need and relevance of collaboration and teamwork. It is becoming more and more evident that great advances are brought about not so much by isolated efforts of individuals, however gifted, as by collaborative efforts of talented, committed, and open-minded persons. The old paradigm of "genius in isolation" for serious, creative achievements is giving way to that of "experts in collaboration." Newton in the seventeenth century and Einstein in the twentieth century did outstanding, creative work almost as isolated individuals. But today the Human Genome Project, the space programme of NASA, etc. owe their success to the well-planned and meticulously executed collaborative efforts of experts from different fields.

Greater awareness and recognition of the power and limitation of science

Another apparent contradiction we find in our world of science today is the growing awareness of both the power and the limitation of science. Never before in the history of humankind have we witnessed the amazing power and capability of modern science. At the same time, never before have scholars exposed the inherent limitedness and inescapable weakness of modern science. Science, indeed, is the creation of humans and seems to share in the limitedness of humans. Gone are the days of scientism when many a megalomaniac scientist assumed that science was the panacea for all problems and the scientific approach the sure path to all success. Today, thanks to several groundbreaking developments in science itself and in the philosophy of science, it is becoming more and more accepted that, although science is mighty and potent, it is not omnipotent. Scientific findings and predictions are reliable but can claim no inerrancy; science has given us the best knowledge, but not perfect knowledge.

I refer to this turnaround as the "humbling experience" of science, wherein science has come to a genuine and realistic self-understanding, recognizing its real power and capabilities, acknowledging its clear limitations and shortcomings. This is a very healthy development that opens the door for greater enrichment. In particular, this opens the door for constructive and creative dialogue between science and other areas like religion.

Shifts from the mechanical philosophy of nature with its mechanistic world to a non-mechanistic, complex world

One of the points emphasized by contemporary philosophy of science is that science is not just a collection of laws, theories. and methodological rules but brings with it a worldview that plays a crucial role in our understanding of the world and in the shaping of our value system. Classical science or Newtonian science was most successful from the seventeenth to the end of the nineteenth centuries and brought along with it a worldview known as the Mechanical Philosophy of Nature (MPN), according to which the universe was a gigantic machine following the rules and principles of mechanics with assured certainty and guaranteed predictability. Practically all the prominent scientists during this period subscribed to this view, to some degree or other, hoping that science would give us sure and absolute knowledge. However, later developments in science, particularly the advent of relativity and the quantum theory, have exposed the deficiency of MPN, revealing the complex, indefinite nature of the universe. The mechanistic view of the universe is not fully given up since even today it persists in many areas of science, particularly in the medical sciences. But the awareness is growing that pure mechanical principles alone cannot capture the complexity and diversity of our universe.

Shift from a deterministic, static, sure, and certain universe to an indeterminate, dynamic, uncertain, and unpredictable world

The mechanistic world was characterized by determinism according to which everything acted in accordance with strict laws of nature. It was also a static world prohibiting any radical change of the system itself, although within the system changes could take place. Furthermore, it claimed to yield sure and certain knowledge about the universe, including human actions. But more recent developments in science like quantum mechanics showed that ours was an indeterminate world; evolution showed that ours was an ever evolving dynamic world; the uncertainty principle of Werner Heisenberg showed that ours was an uncertain and hence unpredictable world. These changes, far from being confined to the academic circles, were recognized as having far-reaching consequences.

Shift from humans as creatures to humans as co-creators: human dignity enhanced

One of the positive results of these developments in science was the enhancement of human dignity. This was particularly true of the developments in the biological sciences, like cloning, the Human Genome Project, genetic engineering, etc., which offered unprecedented powers to scientists to reshape and even alter the nature and function of living beings. In the past, humans were mere creatures living helplessly subject to the creator. With these advances, humans have been elevated to the level of co-creators and collaborators empowered to chart the future course of creation.

Reaffirmation of the unity of the universe

Another far-reaching consequence of the developments in science, particularly since the dawn of the twentieth century, is the reaffirmation of the unity of nature. In the first half of the twentieth century, the different discoveries in particle physics revealed the unity in diversity of the non-living world, since, according to them, the whole material world is made up of the same fundamental particles such as protons, neutrons, electrons, etc. In more recent times, the genome project and related developments show this unity in diversity of the living world. Just as atoms of different material elements are made up of the same fundamental particles, the DNA of different beings is made up of the same kind of nucleotides – A(Adenine), G(Guanine), C(Cytocine), and T(Thymine). Even in the sequencing one can see a remarkable similarity. The genomes of different organisms like yeast, nematode worm, fruit fly, mouse, etc. show remarkable similarity with the human genome. According to some estimates, humans share 99% of DNA with chimps. With the cow the DNA shared is 90%, with the mouse 75%, with yeast about 30%, with E. coli 15%, etc.[1] The human race has crossed the six billion mark some time ago. Despite such large numbers spread over many continents, cultures, and races, humans show a remarkable deeper unity in their biology. It is found that any two individuals differ on the average only in one nucleotide per thousand.

Interconnectedness of the universe – the butterfly effect[2]

Many religious traditions, particularly the Eastern ones, have always upheld the interconnectedness of the myriad beings in the universe. Today this is being confirmed by developments in contemporary science. Quantum theory, particularly as developed in the Copenhagen tradition, strongly endorses this view. The theory of quantum entanglement focuses precisely on this point. The "butterfly effect" in chaos theory, according to which "a butterfly flapping its wings in Hong Kong can affect the course of a tornado in Texas," also carries the same message. The genomic data presented above show how deep the genetic connection is, not only between the different living beings but even more among the billions of human beings. In a real sense we are brothers and sisters, having a real common origin and a common destiny.

This interconnectedness is not confined only to the quantum world or the genomic world. Today we find echoes of it in the world of our everyday experience too. The action of a single person, however insignificant, has the potential to affect the rest of the world. Think of what the action of Bin Laden, who was almost an unknown and unheard-of entity until 11 September 2001, has done to the world. Today it is becoming clear that no person, whatever be his/her status and background, can be taken for granted.

The return of the mystery dimension of the universe

It was hoped, mostly by the opponents of religion and God, that with the emergence and growth of science all mysteries would be demystified, thereby banishing religion and God to the limbo of the superfluous and the superstitious. But this has not happened; on the contrary, the mystery dimension has only reasserted itself, taking up more sophisticated forms. The human genome, more than anything else, reveals the mastery and mystery of creation. It reveals the creator's mastery over creation. At the same time its complexity and intricacy has become a baffling mystery to us humans. The sheer number involved is staggering. It is estimated that in an average human body there are one hundred trillion cells. In the nucleus of each cell there is DNA which contains about 3.1 billion base units or nucleotides, each one of which has over 50 atoms. Now one can figure out how many atoms there are in each cell, how many in the whole human body! And all these trillions and trillions of atoms are arranged in the most orderly manner to make complex life possible. This is for just one human being, and there are over six billion of them walking around on the planet. These facts exposed by contemporary science simply astound us. The only answer Francis Collins, the director of HGP (Human Genome Project), could give to this was "a sense of awe." His sincere and hard-headed science transformed Collins from an atheist in his younger days into a practising Christian today. Speaking of his most important work on HGP, he comments: "I experience a sense of awe at the realization that humanity now knows something only God knew before. It is a deeply moving sensation that helps me appreciate the spiritual side of life" (Cornwell 920). Collins is not the only scientist to have this kind of experience. The well-known British astronomer Allan Sandage too had a very similar experience through his work in contemporary astronomy. Speaking of his turnaround from "almost a practising atheist as a boy" to a believer at fifty, he says: "It was my science that drove me to the conclusion that the world is much more complicated than can be explained by science. It is only through the supernatural that I can understand the mystery of existence" (Begley 47). One can give many other similar cases. Decades ago both Albert Einstein and Werner Heisenberg had voiced similar sentiments.

Scientific developments and the Indian scenario

India is a country that has always accorded great importance to science and technology, and hence these revolutionary developments cannot but have a serious impact on the situation in India. I discuss below briefly some aspects of it.

Some positive aspects

Shift from a backward, primitive, un-technical, underdeveloped country to a technologically advanced, vibrant nation

In the past, particularly during the colonial and initial parts of the postcolonial times, the stereotype perception of India was that of a third-world, third-rate country teeming with poverty, hunger, illiteracy, and superstition – all leading to an underdeveloped nation of no serious significance. But today this questionable perception has taken a total turnaround. According to reliable reports, today:

- India has the second-largest pool of scientists and engineers in the world
- India is among the top ten most industrialized nations
- India is ranked the sixth country in the world in terms of satellite launches
- India is the only country, other than the US and Japan, to have built a supercomputer indigenously
- India is ranked second in the software industry
- Four of ten Silicon Valley startups are run by Indians
- There are 3.22 million Indians in America. Their involvement in the field of science and technology in the US is impressive. For instance, 38% of doctors in America are Indians; 12% of scientists in America are Indians; 36% of NASA employees are Indians; 34% of Microsoft employees are Indians; 28% of IBM employees are Indians; 17% of Intel employees are Indians; and 13% of Xerox employees are Indians.

According to one expert, "India finally has started acting as the technology superpower in the 'new world' where countries become superpower by virtue of technical strength and capability and not colonial wealth!" (www. frdavis.in/2013/12/value-education-prospects-and-challenges.html). Albert Einstein has the following to say about India: "We owe a lot to the Indians, who taught us how to count, without which no worthwhile scientific discovery could have been made" (cited in *Vedic Revelations* https://books.google.co.in/books?id=IUr-Pl-XK8kC&printsec=frontcover&redir_esc=y#v=onepage&q=einstein&f=false).

Shift from a poverty-stricken, impoverished colony to a potential economic giant

For centuries India was looked upon as a poverty-stricken, impoverished state colonized by foreign masters. But today this situation has changed significantly.

- According to the World Bank, India is now listed among the top 10 economies of the world
- India has become an attractive centre for investors. As Bertie Ahern, former Prime Minister of Ireland, says, "India's unprecedented economic growth over the past decade makes it an attractive prospect for companies seeking new markets for their products and services" (www.business-standard.com/article/technology/ireland-wants-high-tech-india-as-partner-106011801097_1.html).

Shift from a land of diseases, epidemics, malnutrition, and other health hazards to an attractive centre of medical tourism

In the past India was often depicted as a land of high health hazards, abounding in diseases and epidemics. Foreigners, particularly from the developed, affluent nations, hesitated to come to India. But today the scenario has changed dramatically. India has become an attractive spot for medical tourism, second only to Thailand. Of course, India has always been noted for Ayurvedic medicines, massage treatments, yogic meditation techniques, and other ancient medical traditions. However, today she is at the forefront of modern medicine as well, attracting huge numbers to her state-of-the-art medical facilities in various parts of the nation. The records in these world-class medical institutions are impressive. For instance, Indian specialists have performed over 500,000 complex surgeries such as cardio-thoracic, neurological, and cancer-related surgeries. The success attained in these surgeries is on par with international standards. The success of cardiac bypass in India is 98.7%, higher than that of the US. The success rate of renal transplants is also 95%. Furthermore, India is home to the largest number of pharmaceutical plants (61) approved by USFDA in any single country outside the US.

The health condition of the ordinary citizens also has seen remarkable improvement. For instance, the average life expectancy in India has increased considerably. Life expectancy in 1970 was 49, in 1990 it rose to 59 and in 2006 it reached 64. On the other hand, death rate declined from

48.6 per 1,000 in the 1910–20 period to just 15 per 1,000 in the 1970s, and improved thereafter, reaching only 10 per 1,000 by 1990, a rate that held steady through the mid-1990s.

Increasing food production and move towards self-sufficiency

With regard to the production of food also India has made great strides. The statistics in this regard speak for themselves. For instance, the total area under the high-yielding-varieties programme was a negligible 19,000 km² in the financial year 1960. However, since then growth has been spectacular, increasing to nearly 154,000 km² by the financial year 1970, 431,000 km² by the financial year 1980, and 639,000 km² by 1990.

Increasing urbanization

Steadily increasing urbanization is another consequence of the developments in science and technology, although, compared to many other countries, India's pace of urbanization has been slow. At the moment, India is among the countries of low level of urbanization. The number of urban agglomerations/towns has grown from 1827 in 1901 to 5161 in 2001. The total population residing in urban areas has increased from 2.58 crores in 1901 to 28.53 crores in 2001. Only 28% of the population is living in urban areas according to the 2001 census.

Beauty contest

Even in areas like beauty contests, India has made her mark. Six Indian women have won the Miss Universe/Miss World titles over the past ten years. When we realize that no other country has performed so well, we begin to appreciate this achievement as something remarkable.[3]

Some negative aspects

Shift from self-reliance to techno-reliance

Indian culture and tradition were noted for its simplicity, practicality, closeness to nature, and self-sufficiency. Mahatma Gandhi crystallized these effective, time-tested traits in his concepts of *swadeshi* and *swaraj*, in which the aspect of self-sufficiency and self-reliance are fundamental. In his own words, "My idea of village *swaraj* is that it is a complete republic, independent of its neighbours for its own vital wants, and yet interdependent for many others in which dependence is a necessity." This would mean that the epicentre of Indian culture and tradition are the villages. In Gandhi's vision, "The true India is to be found not in its few cities, but in its seven hundred thousand villages. If the villages perish, India will perish too." Furthermore, this would mean that we focus on production by the masses rather than mass production. In Gandhi's view, mass production is only concerned with the product, whereas production by the masses is concerned with the product, the producers, and the process.

In contemporary India, thanks to the explosive growth of science and technology, this tradition is fading away, and a culture of techno-reliance is fast spreading. People become so dependent on machines that they cannot manage if the machine system breaks down, as evidenced almost daily when we are afflicted with electricity power cuts. This change has serious consequences. The simplicity of life is lost; the use of hands becomes more and more rare; unwanted and unnecessary dependence sets in. Accompanying this disregard for nature and what is natural is the mechanical mentality which looks upon fellow humans as machines and which wants to analyze their life and behaviour mechanistically.

Economic and cultural exploitation

Another negative aspect of the technological development is that it can easily become an effective tool of exploitation in the hands of profit-hungry multinationals and technocrats. In fact, it has already become so in many cases. This has come about because technology requires highly expensive infrastructure, which only very few can afford. Again, the nations and companies which are already ahead and well-established have a decisive advantage. Also, today technology and regulations are such that those ahead can claim a monopoly and prevent others from coming up. For instance, today the genetic technology can produce terminator seeds that do not allow new seeds to be produced from them, forcing the farmers to purchase fresh seeds each time they want to cultivate. Since only the original company with a monopoly and copyright can produce and sell these seeds, the farmers are forced to buy the seeds from the parent company, often at an unjustly high price.

A number of instances of this kind of economic colonization and exploitation have been reported and documented. It has been reported that these companies with very good capital base, particularly in India because of the attractive rupee foreign exchange rate, first buy up all possible local competition by offering a highly attractive and even disproportionate price. For instance, it was reported some years ago that in Brazil the Monsanto Company – a US multinational agriproduct company headquartered in St. Louis, Missouri – spent more than $1 billion to buy out 60% of all the seed companies in just two years. In India it bought major holdings in the largest seed company. Once this is done, these multinational companies introduce their own product along with their production technology and monopolize the field. It may be noted that the seeds produced by these genetically engineered products either do not germinate or need for germination certain specific chemicals developed by these companies. This means that once farmers begin to use this product, they are permanently dependent on the companies, leaving these multinational giants full freedom to exploit the local people.

In the case of Monsanto, it was further reported that in 1999 the company was engaged in carrying out field tests of its genetically engineered crops in forty locations in India. More specifically, the trials were on Monsanto's genetically engineered "Bollgard" cotton, or Bt-cotton, produced by engineering genes from a bacterium into the cotton plant in order to enable it to produce its own pesticide. Critics point out that Bt-cotton is not pest-resistant but a pesticide-producing plant: "The severe ecological risks of crops genetically engineered to produce toxins include the threat posed to beneficial species such as birds, bees, butterflies and beetles which are necessary for pollination and for pest control through pest-predator balance" (Siva 100–1). Monsanto claimed 93% success for its product but failed to point out that this referred to agronomic performance only and did not take into account ecological and other concerns. The principal complaint in this context is that these huge companies use the less developed nations for testing out their controversial products and the helpless farmers for marketing them at monopoly-driven prices.

There is not only economic exploitation but also cultural exploitation, often introduced in a subtle, innocuous manner as part of entertainment, advertisements, etc. Cultures and cultural values have specific contexts, and once detached from these contexts and introduced into another culture, they can become not only unhelpful but even harmful. This is already happening in India today. For instance, family values are taking a strong beating in many parts of India, particularly in the urban areas. It has been pointed out that one of the sources of this tragedy is the invasion of the entertainment media from certain countries where family values have lost their centrality.

Tension between material prosperity and spiritual poverty

Progress in science and technology is often accompanied by material prosperity, particularly in the technologically advanced countries and groups. More food, better education, improved health, more leisure, longer lifespan, etc. are the usual indicators of this prosperity. However, it has been found that this material prosperity is not accompanied by any spiritual or inner growth. In fact, in some cases one can see a concurrent decline in the world of the spiritual. Commenting on this asymmetry in the growth of the material and spiritual worlds, Louis de Broglie lamented: "Our enlarged body clamours for an addition to the spirit" (Wilber 121). "Now, in this extremely enlarged body, the spirit remains what it was, too small now to fill it, too feeble to direct it," he continued,

"Let us add that this increased body awaits a supplement of the soul and that the mechanism demands a mysticism" (Wilber 122). In his view,

> Humanity groans half-crushed under the weight of the advances that it has made. It does not know sufficiently that its future depends on itself. It is for it, above all, to make up its mind if it wishes to continue to live.
>
> (Wilber 122)

This imbalance between material affluence and spiritual hunger cannot be brushed aside as something insignificant or irrelevant. This is a serious matter demanding our careful attention, since this mismatch exhibits itself in many ways, like decline in the law and order situation, increase in the cases of depression and suicide, etc. Our contemporary society in many parts of the world, including India, is afflicted by this unhealthy situation.

Widening of the gulf between the haves and the have-nots

Another unhealthy outcome of the explosive growth in science and technology is the widening gulf between the rich and the poor. The economic growth, in many ways, has been one-sided, giving rise to a deformed monster rather than a well-balanced organism. The rich seem to be getting richer and richer, while the poor are becoming poorer and poorer. Nowhere is this anomaly more conspicuous than in India, which still has the world's largest number of poor people in a single country. Of its more than one billion inhabitants, an estimated over 350–400 million are below the poverty line, 75% of them in the rural areas. The level of illiteracy is alarmingly high, particularly among women. More than 40% of India's population is illiterate, with women, tribals, and scheduled castes particularly affected. With regard to infant mortality, the statistics are equally dismal. India is ranked fifty-third in the world with 55 deaths per 1000 births, and 78.6 deaths per 1000 births in the case of children under 5.

Rise in environmental degradation and pollution level

Another unfortunate victim of all these developments triggered by an unprecedented advance in science and technology is Mother Nature, particularly the environment. Much has been written on this problem under the heads of deforestation, acid rain, greenhouse effect, ozone depletion, etc. Pollution in all its virulent forms – air pollution, water pollution, and land pollution – has been inflicting serious health problems on India. A number of Indian cities have been listed among the most polluted ones. Many steps have been taken by governmental and non-governmental agencies, but much more needs to be done.

Some serious consequences of these developments

The cultural confusion

There is no doubt that the astounding developments in science and technology have played a major role in bringing about the changes and shifts, with their accompanying blessings and curses, as discussed above. No doubt several other factors also have their share of responsibility. But since the capability of science to strike immediate results is far greater than that of many other influencing agents, the major responsibility rests with it. Moreover, since science's forward march is an ongoing one with ever increasing pace, one can expect more surprises on the way.

Perhaps one of the immediate consequences is the cultural confusion these developments are giving rise to, presenting formidable challenges to various sections of our society. It is clear that many of the items in this package are good and are to be encouraged. For instance, many aspects of globalization, with its broadening of vision and opening up of new vistas, are good. But the challenge is how to safeguard many legitimate local needs and concerns in this process of globalization so that the individuality and uniqueness of local cultures is not sacrificed. So often we are saddened by the spectre of many abandoning their Indian cultural values in a mad chase after certain Western fads.

The challenge is not something that will go away with time. In fact, it will only become stronger and more formidable since there is no turning back of this process. Nor is the traditionalists' solution of "retreat to the good-old days" a healthy response to this challenge. Ours is an evolving world; evolution is not only a law of life, it is also a law of the universe. As Teilhard de Chardin and other thinkers have passionately and persuasively argued, no power on earth can stop this process of ongoing evolution. This world of ours will have to learn to deal with this situation constructively and creatively. Therein lies the path to future success.

How to face this challenge?

At the very outset we need to admit that this challenge is only natural and to be expected since culture by nature is a dynamic, growing/evolving phenomenon. As the "Theme Overview" of this seminar puts it, culture can be looked upon as "the sum total of all the devices and methods that a society has at its disposal to control, direct, and modify the material conditions of its existence." In the case of such a complex, dynamic phenomenon, changes and challenges are to be welcomed as part of the growth process. History tells us that the Indian culture is one of the most ancient and experienced ones, having had to face the onrush and impact of many outside alien influences, even destructive ones. India has faced them successfully and creatively in the past. No culture or outside influence that has come into close contact with India has remained unaffected by some aspects of her unique traits. The well-known American writer Mark Twain has remarked: "India is the cradle of the human race, the birthplace of human speech, the mother of history, the grandmother of legend and the great grandmother of tradition."

The present challenge seems to be rather formidable, having many apparently conflicting strands defying any easy integration. We are called upon to do due justice to the different strands without undermining the essential elements of our own culture. Among other things:

- This would mean the challenging task of keeping our technological edge that is essential for growth, without sacrificing the human face
- It would mean keeping the path of material prosperity open without marginalizing those below the poverty level
- It would mean developing medical treatment, without ignoring the basic medical facilities for the ordinary people
- It would mean that India's economic plans should focus *not only on profit-making but also profit-sharing*, particularly with the less-privileged ones
- It would mean that science and technology be put primarily at the service of meeting the basic needs of humans, rather than making lucrative cosmetic items for the rich and famous
- It would mean using the almost inexhaustible nuclear power for useful, peaceful programmes, rather than for the production of more and more sophisticated weapons of mass destruction
- It would mean transforming the national and multinational corporations from profit-hungry organizations to service-sensitive ones
- All cultural, value-based restructuring is required in the light of these developments, giving rise to a new culture, a new world order, a new India

Science-spirituality dialogue and the new world order

In this final section I wish to point out that a creative and constructive inter-action of science and spirituality is necessary for this cultural restructuring. This collaboration is only natural since science is today an integral part of any culture, particularly our Indian culture, and spirituality has always been a central part of it, although it is expressed in different ways.

This all-important process of value-based cultural restructuring is a long and complex process requiring judicious and skillful intermingling of many elements. This certainly involves adapting and integrating the emerging new trends into the cultural setting of India. Not all new trends can be assimilated wholesale; nor should all the traditional values be thrown overboard. Identification of the new items to be admitted and of old ones to be abandoned will have to be done with the utmost care and sensitivity.

Certain traditional values to be reemphasized

This process also involves the reemphasizing of many of our traditional, time-tested values and principles. For instance, India was noted for her positive appreciation of other cultural values. History tells us that we accepted other cultures with a welcoming attitude. This spirit has to be preserved and further developed. Again, India was noted for her spirit of tolerance. In fact, many serious scholars have shown that Hinduism is the most tolerant among the major religions of the world. In the process of cultural restructuring, this spirit will be of immense assistance. The spirit of *nishkamakarma*, selfless service, was another value very much emphasized by our Indian religious traditions. In today's profit-hungry, success-oriented world driven by cutthroat competition and rivalry, this principle should play a central role. The spirit of the *mahatma*, large soul, is another treasured value in our rich tradition. This spirit too needs to be reemphasized. I leave the experts to extend this list further.

New values to be developed

This process also involves developing new values in the light of the new developments – a task demanding well-informed, discerning, creative, innovative, and daring persons. Here also a constructive interplay between scientific and spiritual values can be of immense help. To suggest a few possibilities, along with the scientific values of efficiency and expediency, concern and care for others are to be cultivated. Again, along with the scientific values of exactness and certainty, room will have to be provided for a certain imperfection and uncertainty. Furthermore, along with the scientific values of predictability and self-reliance, provision will have to be made for a certain unpredictability and reliance on other sources, particularly on divine providence.

Conclusion

The challenge is clear, but the line of response awaits further clarity. But one thing is non-controversial: neither science alone, wherever its success may take us, is competent to meet this challenge, nor is spirituality left alone, whatever be its resources, competent to deal adequately with this situation. On the other hand, together they can meet the challenge more effectively, creatively, and constructively. A serious, well-planned, and skillfully executed dialogue between modern science and spirituality should be one of principal players in the creation of a new world order for the betterment of humans and the cosmos.

Notes

1 For details see Michio Kaku, *Visions* (Oxford: Oxford University Press, 1998), pp. 152–153. Please note that the percentages given are only approximate, since different researchers give slightly different figures.
2 About eighty years later, as early as 1963, Edward N. Lorenz, using Poincaré's mathematics, described a simple mathematical model of a weather system that was made up of three linked nonlinear differential equations that showed rates of change in temperature and wind speed. Some surprising results showed complex behaviour from supposedly simple equations; also, the behaviour of the system of equations was sensitively dependent on the initial conditions of the mathematical model. He spelled out the implications of his discovery, implying that if there were any errors in observing the initial state of the system (which is inevitable in any real system), prediction as to a future state of the system was impossible. Lorenz labelled these systems that exhibited sensitive dependence on initial conditions as having the "butterfly effect": this unique name came from the proposition that a butterfly flapping its wings in Hong Kong can affect the course of a tornado in Texas. This has become the emblem of chaos theory, following James Gleik.
3 Some of the countries that come close to India are Venezuela, which has won Miss World five times, followed by UK, which did so four times.

Works cited

Begley, Sharon. "Science Finds God." *Newsweek.* 20 July 1998.
Cornwell, John. "Scientists Playing God." *The Tablet.* 8 July, 2000.
Siva, Vandana. "Monsanto's Genetic Engineering Trials in India Are Dangerous and Anti-Democratic." *Third World Resurgence.* Vol.100 (December 98–January 99).
Wilber, Ken. *Quantum Questions.* Boulder, CO: Shambhala, 1984.

6

KNOWLEDGE AND SCIENCE IN THE CONTEXT OF INDIAN LANGUAGES

Probal Dasgupta

What geometry was in the ancient Greek context, grammar was in the ancient Indian context. If some of us begin to look at the approach to knowledge and science that developed around the notions of grammar in ancient Indian thinking, this transformation we undergo opens up for us the field of inquiry in ways that are not exclusively Paninian. The present intervention contextualizes these debates in the context of generative reappropriation of ancient grammatical theorizing.

Our standard access to ancient Indian grammatical work is through contemporary linguistics and specifically through its grammatical theorizing. There is an imbalance here that needs to be corrected. Simply introducing a "socio-linguistic" tilt is not enough, for the implied binary between formal work and social inquiry is inappropriate. The generative reappropriation of ancient grammatical theorizing is a context that we need to overcome even as we respond to its imperatives. The social needs to be restored within grammatical inquiry itself; the attempt to introduce a sociological supplement misses the point. Hence the format of the present intervention.

Whatever stand scholars might choose to take in the context of the orientalism debates, the study of ancient Indian linguistics calls for particularly sensitive handling of methodological issues for at least two reasons. One of these is familiar – and will drive most of the discourse in the present exposition: contemporary generative grammar works with presuppositions and procedures very close to assumptions that were standard in ancient Indian grammatical thinking. Another, less familiar reason has to do with the role of classical Indian categories in the normative grammars of modern languages in South Asia.

To see this second point more clearly, it is useful to compare grammatical and medical discourse. Ayurvedic wares have a reasonable market, but Ayurveda's map of the human body does not underwrite the way the South Asian middle class speaks of metabolism and illness. In contrast, standard discussions in Kannada, Bangla, or Gujarati of such matters as pronouns, nominal inflection, or verb-actant bonds routinely invoke concepts like *sar-vanaama*, *vibhakti*, and *kaaraka*.[1]

In other words, work done today to advance our understanding of the Indian grammatical tradition needs to forge links with the way in which the tradition's concepts are recycled in relation to modern South Asian languages. Besides, many scholars have long felt a similar need to associate such inquiry with the contemporary generative retooling and expansion of the formal grammatical project that has come down to us from classical Indian work. These areas of interdisciplinary contact have not yet been fashioned at a sustainable level.

This omission may not be all that difficult to understand. One probably tends to regard as an unsolved problem the entire task of bringing classical Indian traditions as a whole to bear on modern pursuits. But neither normative grammar writing in modern South Asian languages nor the global consolidation of generative grammar can make serious progress if these bridges are left unbuilt. In that sense, classical Indian linguistics is a bit of a special case and calls for an immediate and principled response.

The point cannot be to press for the production of books for which neither authors nor audiences are available. Work will proceed at its natural pace and cannot be hastened. Fashioning a principled response involves changing how one thinks of the work and what one does with the existing literature. Specifically, we need to outgrow the objectifying style in Sanskrit grammatical studies (orientalism is only a special case of this) and to bring the necessary inter-subjectivity into play in the concrete functioning of this field's academic apparatus.

To get there, one must of course begin by engaging the agency of the workers themselves. Therefore ensuring access and inspectability certainly deserves the priority it has received in scholarly work on the Indian grammatical tradition. Only if workers in the field have effective access to relevant materials can inter-subjective calibrations begin to take off. We must now respond, however, to the fact that this is no longer the principal challenge to be faced: effective access to the formal and technical core of Panini's system now exists. With the publication of milestones such as Cardona (1988), Joshi and Bhate (1984), Joshi and Laddu (1983), Katre (1989), and Subrahmanyam (1999) for Paninian studies, and given the availability and undiminished usability of Pillai (1971) and Subramania Iyer (1969) for the Bhartriharian sequel, a certain threshold has been reached, calling for a change in priorities.

It is inevitable, and desirable, that there should be debates at a time like this. The exposition here represents a viewpoint that sees Bhartrihari as the culmination of the Paninian tradition and seeks to incorporate his approach into contemporary work. Hard-core Paninians, who prioritize the formalist view of language, treat Bhartrihari's *Vakyapadiya* (or to transcribe it more precisely, his *Vaakyapadiiya*) as an optional discourse-based supplement to the sentence-grammatical endeavour and may be expected to wish to preserve the autonomous purity of sentence grammar. Votaries of formalist versions of generative grammar who have some awareness of their role as today's Paninians will also, one hopes, provide some articulate opposition to the substantivist approach presented here.

The issues at stake, as will become apparent in the course of this exposition, are partly methodological and partly empirical. The composition of this basket of issues is bound to move the debate from the purely Sanskritic terrain on to the serious formal study of language more generally. As the neo-Paninian formalists in generative grammar begins to take the empirical points on board, the debate will cease to be confined to ancient Indian contributions to Sanskrit studies and will spill over into the serious study of knowledge of language in general. These are some considerations that motivate the choices and emphases in this exposition.

In the present intervention, Formalism and Substantivism, as two distinguishable approaches to the study of language, are conceptualized as follows. The formalist approach focuses on the grammatical rule as the austere formal object of rigorous statement. It seeks to maximize the economy of grammatical rule formulations. All other methodological decisions flow from the primacy of the rule of grammar. Formalist methodology seeks to associate one formal object with one rule bi– uniquely and aggregates rules to establish as unitary a system as possible.

The substantivist approach regards the cycle from sentence composition through speaking, hearing, and understanding to fresh composition as the rich substantive domain of grammatical inquiry embedded in the context of discourse. It seeks a maximally transparent and economical account of this cycle within which rules of grammar and other descriptive devices are to be seriously conceptualized, going beyond abbreviations that may work at a first approximation level but are not sustainable. Substantivist methodology uses translation to associate each formal object with several semiotic systems.

The partly contradictory demands of the twin imperatives – the formalist imperative of writing a tight grammar and the substantivist imperative of providing a coherent account of discourse – were noticed quite early in ancient India. Around the time that Panini's grammar was codified, the scholar Vyadi a.k.a. Dakshayana, traditionally identified as Daaksxiiputra, Panini's maternal uncle, authored a major commentary called the "Samgraha". Its text has not come down to us, but references to it allow us to reconstruct its scope (Subrahmanyam 1999, 21). Bhartrihari's much later work *Vaakyapadiiya* rearticulates and codifies the project initiated in Dakshayana's early substantivist supplement to Panini's mostly formalistic grammar.

In the context of the generative re-run and amplification of the ancient Indian grammatical research programme in our times, substantivism becomes a matter of bridge-building between grammatical theory and the study of the actual use of language. We can find the resources for such bridge-building in Bhartrihari's reconfiguration of Panini's apparatus. To briefly outline the itinerary of the present study, we revisit Bhartrihari's proposals, showing that they extend and amplify the logic of Panini's grammar itself, and we explore certain ways in which they can be rendered operational in our context. This inquiry leads us to focus on certain points at which formal grammar most obviously interfaces with pragmatics.

The need to contextualize formal grammar in a social and philosophical matrix was seen at the very inception of the formal grammar enterprise. It is generally recognized that the enterprise of formal grammar writing in general was launched by ancient Indian grammarians working at the intellectual heart of its culture and that Panini's *Ashtadhyayi*, read as an unsupplemented formal apparatus, represents the summation of what unfettered formal grammar can accomplish.

Scholars in our period need to revisit the fact that at that very moment formal grammar had received a contextualization, which we have lost and which we will have to retrieve (even though certain texts are gone forever) both for an understanding of ancient Indian work and for our purposes today. Classical Indian thinking was particularly focused on concrete characterizations of problems and solutions. For us to read the *Ashtadhyayi* as an unsupplemented, bridgeless island of formal abstraction does not square with our overall sense of Panini's matrix.

The standard point of departure for grammatical philosophy is Bhartrihari's *Vaakyapadiiya*. The textual, hermeneutic link between the *Vaakyapadiiya* and Panini's oeuvre was through the *Samgraha*, which, alas, is lost. Dakshayana's *Samgraha* built what seems to have been a contextualization of formal grammar in its substantive matrix; whether we believe or disbelieve the tradition on the kinship between Dakshayana and Panini, we must recognize that such a kinship attribution is one way of saying that the two oeuvres belong together. At the end of the Paninian odyssey through Katyayana's *vaarttika*s and Patañjali's *Mahaabhaasxya*, it is Bhartrihari's *Vaakyapadiiya* that renews and codifies the *Samgraha's* contextualizing enterprise.

The view that we now face the task of renewing his renewal, with generative grammar as our period's counterpart to Panini, and even the term substantivism as a label for the enterprise that needs renewing, were first presented in 1989 (Dasgupta 1989). Before we turn to the work of renewing it, however, we must first make sense of substantivism as it first appears in ancient Indian inquiry. The bulk of the present exposition is devoted to the labour of this sense-making.

The protracted ancient debate at the end of which Panini's work became possible pitted a view of the word (*s'abda*) as eternal, originary, and pre-derivational (*nitya*) against the view that every word was derivable (*kaarya*).

Panini accepted the logic of the project that sought to derive nominal and adjectival words in the lexicon from verbal roots by applying affixation. But he did this within limits.

The derivational logic made him derive the word *rikta* "empty, vacant," for instance, by applying the suffix *Kta* to the root *ric* "to evacuate." Only the root and the suffix would count as basic, then. However, in the case of certain nouns like *as'va* "horse," Panini did not push the derivational logic all the way; he stopped short of deriving words like these from a root-suffix combination. He thus treated certain nouns as basic, contrary to the extreme "derived-word" doctrine that would have required the postulation of a theoretical root-suffix pair *as'-va* for purely formal reasons.

That extreme view would allow that individuals might have forgotten other uses for that root and that suffix but would insist that the existence of the noun made no sense unless it was derived from a root-suffix combination. What is sometimes seen as Panini's pragmatism in this matter represents a concession to the "originary-word" doctrine and enables it, with crucial support from Patañjali, to make the comeback that is codified in Bhartrihari's *Vaakyapadiiya*.

Panini's partly derivationalist, partly originarist practice as a grammar writer represents the same early phase of the tradition as Dakshayana's apparently equally balanced presentation in his *Samgraha*. We know this from Patañjali's succinct description of that exposition, which is worth citing.

In his *Mahaabhaasxya*, one of the few sources of information about the scope and content of the *Samgraha*, Patañjali writes:[2]

> Is the word originary or derivative? In the Samxgraha the question whether the word is originary or derivative has been mainly examined. There the defects and the uses (of these views) are explained. There it has been decided that grammar is essential either way, that is, whether the word is originary or derivative.

Patañjali's own conception of the agenda of grammar tilts the balance in favour of the originarist view. In the context of finding social coordinates for the labours of the grammarian, Patañjali writes:[3]

> One who wants to use a pot goes to a potter's house and says: 'Make a pot! I will use it.' One who wants to use words will not similarly go to a grammarian's house and say: 'Make words! I will use (them).' (People) use words taking meanings (to be expressed). If, then, society (the world) is the authority in the matter of words, what does the science (of grammar) do?

Of course Patañjali does list the standard reasons for continuing the enterprise[4]; but his reading of the tradition locates the centre of gravity of the normative project in the native speaker's judgments. This construal is what makes possible the originarist comeback codified by Bhartrihari. So we need to slow down and look at Patañjali's move – or at what Patañjali is, not unjustifiably, asking us in his commentary on Panini to regard as Panini's move – with some care.

The pre-Paninian debate was about whether lexical items, sabdas, were derived in general (the etymology-maximizing position) or originary in general (the etymology-dismissing position). But Panini's machinery sets out to derive within the grammar not just the raw lexical sabda but the cooked syntactic pada, the fully inflected word as it is deployed in relation to its neighbours in the sentence.

In other words, Panini takes a *kaarya-pada* position that recasts the old *nitya-s'abda* vs *kaarya-s'abda* debate. He proposes that a syntactic word, a *pada*, must always have its endings licensed by the grammar and thus must count as derived by the apparatus. The stand he takes on the lexicon's inflection-subtracted bare stems or *sabdas* is that many but not all sabdas are kaarya, i.e., derivable within his system. To revisit examples given earlier, he derives *rikta* "empty" from a root-suffix combination; but *as'va* "horse" counts as an originary, underived lexical item.

Methodologically, the Paninian project can thus be construed as setting itself the goal of formalizing coherently and completely what might be expected to be an idealized native speaker's judgments pertaining to the well-formedness of words and to inter-word relatedness. This construal – that the project is to derive all and only those cases that are rationally decomposable into roots and affixes – helps make sense of the coverage decisions, including the omission of items like *as"va* from the scope of the etymologizing project.

If we accept such a reading of Panini's enterprise, it becomes natural to accept as a fair characterization of the entire Paninian tradition Patañjali's view that it is the native speaker's knowledge that initially validates words. On that assumption, the grammar, by deriving the inflected words in every sentence (and by providing lexical item derivations in many cases), offers a second, formal validation of words by way of articulating the logic of the native speaker's knowledge of the system.

The picture of the basic moves now comes out as follows. Panini, who establishes and mostly implements the project, and Katyayana, who completes the implementation, takes it that the grammarian's task is to provide a formal account of all valid (siddha) words. They formalize the work of *saadhana* (validation) as a *prakriyaa*, an operational derivation, on the basis of a methodology that seeks to derive all and only those words that are rationally decomposable. In their work, simplicity is a matter of packaging this derivational system in the simplest form possible, a point we need not labour for this readership.

Once they complete the initial project, Patañjali is able to add a fresh question – given the decision that there are certain words that this *prakriyaa* is supposed to derive and validate as *siddha,* he raises the question of what extra-systemic source these words are getting their initial legitimacy from. (To this question his answer of course is social currency among native speakers.) Patañjali in effect adopts a double validation approach, noting the social and textual authority for a word as an independent source against which formal derivability must cross-check to achieve a concrete theory anchored in speech and reference.

Historically, Patañjali's work triggers Bhartrihari's sequel, though this sequel crucially bases its methods on Dakshayana's much earlier intervention. Methodologically, double validation is the point at which formalism begins to count as just an approximation to the richer substantivist account. For the formalist, there is only one exit from language, at the end of the formal route through the derivation itself. But the substantivist allows for the discursive agency of the speaker and the listener, and pushes the multiplicity of formal grammatical modules towards a notion of economy based on transparency rather than on the formal parsimony of austere lexical and grammatical statements.

This contrast is important in the context of linguistic research in our time; linguists are right now going through a transition from formalist methods based on a first reading of Panini's grammar and Panini-type sequels to substantivist methods of the sort first worked out by Bhartrihari. But the contrast matters also in the context of trying to understand the larger transition in Indian history from the Vedic and classical period to the mediaeval formation usually described in terms of bhakti and Sufism, a topic to which we briefly return later in this exposition. For the moment, we concentrate on the broadening of the scope of grammar in Bhartrihari.

Patañjali initiates what can with hindsight be conceptualized as substantivist grammar's fundamental problem – that of the grammatical and discursive double validation of the syntactically deployed word, the *pada*. Bhartrihari develops and addresses that fundamental problem.

In Patañjali, the *pada*'s independent reality perceived by standard speakers and seen in standard-setting texts is a site of inquiry that can lead to theoretical concreteness. But his approach recognizes only the derivation as a locus for rigorous work. The social reality checking is for him only a public filter that has no operational impact on the formal derivations. Bhartrihari expands the domain of the formal so that the *pada* travels across speaker's comprehensions of productions to end up with an operational result. Thus the formal for him includes exits from the sentence that are not routed through the derivation.

How does Bhartrihari attain this result? He does it by choosing as the grammarian's epistemic vantage point the infinitely responsive listener's position rather than that of a speaker pointlessly churning out endlessly many sentences. His ideal listener who can also hear non-language makes allowances for non-standard dialect speakers and proto-articulate toddlers but positions himself at a standard adult approach to linguistic validity. Translation is the procedure that underwrites his auxiliary operation of making allowances. Bhartrihari's emphasis on the cognitive goes so far as to explore the heterogeneous but connected distribution of concrete knowledge over the social network. In terms of method, Bhartrihari gives us a multi-validated meta-grammatical object by filtering grammar formally through the pragmatic agency of the ideal hearer.

Is Bhartrihari's object of inquiry the sentence rather than the word, as a first reading of his text might suggest? If we look at the notion of the *sphota*, the widely known heart of Bhartrihari's conceptual enterprise, his method of inquiry gives us a multilevel *sphota*. This is made possible by the recourse to decisions by the listener. This recourse is a versatile device and can take in its stride the *varna* "sound," the *pada* "syntactically deployed word," and the *vaakya* "sentence." But his architecture builds in the obvious asymmetries. The *varna* is too small to count as his basic unit. The *vaakya* is too big for the purpose. The *pada* is just right, and is also – in the perspective he inherits from Patañjali and encodes in the formal apparatus – both derivationally validated and socially co-validated.

In other words, on our reading, Bhartrihari's doctrine that the *vaakya* is the real unit of language is best seen as a matter of connectivity between *pada* and *pada* in the syntagmatic special case. If this reading can be sustained, we can take him to be saying that the texted *pada* is the prototypical reality that speakers and listeners invoke and that the grammarian needs to access.

This reading is confirmed when we look at the planes of synthesis Bhartrihari postulates, which too are not sentence-driven. His well-known planes *vaikharii*, *madhyamaa*, and *pashyantii* are mutually articulated at the level of the perspective within which the speaker and the hearer attain a shared understanding of the context of speech as well as the content.[5]

This take on language says that the perspective shared by speakers and hearers underwrites the discourse. The particular flaws of what is said and understood, from *pashyantii* through *madhyamaa* to *vaikharii*, are the traffic, not the roadmap. To the extent that Bhartrihari's work can be validly interpreted as anchored at the level of the perspective shared by the hearer with the speaker, it cannot give formal primacy to the *vaakya*, which is just one of the units on the operative planes. No theory of perspective can afford to get fixated on the sentence. One cannot aim for anything less than the discourse.

Of course, as a technically and operationally *pada*-based view, Bhartrihari's approach gives full play to the distinctively Paninian *kaaraka*-theoretic view of inter-word connectivity. That view certainly does treat the clause as a salient domain in the operation. But we cannot leave it at that.

Clause structure is concerned with the dependency relation between a verb and its dependents (specifically, its actants) that play various *kaaraka* roles. It is the formal task of grammar in the narrow sense to describe dependencies in terms of the asymmetry between the verb and its *kaarakas*. But meta-grammar must perform the substantive task of holding all connectivities together. Dependency is only one special case of connectivity. An excessive focus on the formalization of dependency tends to short-circuit meta-grammar's substantivist work. That technical short circuit, often accepted as a pedagogic starting point, is the formalist exercise of unsupplemented formal grammar.

The technical first approximation exercise is methodologically incomplete but produces the illusory impression of giving us a complete scientific discipline on a formal platter. Linguistics is all about outgrowing that illusion.

What Bhartrihari's overall anchoring at the level of sharing between speaker and hearer does is link what is said to what is not said, formally placing the whole set of entities and inter-entity links within language. He is thus visualizing the domain of inquiry from a privileged perspective locatable in language and yet not confined to language alone. Such a discourse perspective, necessarily multiple and translational in its concretely visualizable implementations, conceptually and methodologically calls for a substantivist apparatus in linguistics. Such an apparatus enables its users to make sense of relations between systems in terms of multiple coding and translation bridgework.

Just how does Bhartrihari's version of substantivism link the said to the unsaid? In his work, what can be known can also be taught. When a craftsman teaches his art to an apprentice, he conveys much of his teaching through "showing" rather than through "saying" in sensu stricto. Bhartrihari makes the conceptually crucial move of construing the continuity between verbal and non-verbal teaching in terms of effability.[6] This has the consequence that in his work the unification of language and non-language takes place in rather than outside language. Language is not a special case of reality. Rather, reality must invariably be said to instantiate language, for formal reasons having to do with what is sayable.

In Bhartrihari, anything that is real counts as transmissible (for all realities are processes) and therefore must in principle be teachable and therefore sayable, even if some of the saying would involve fashioning words that humans have no use for and thus do not in fact give social shape to. The possible words that humans do not use are, in Bhartrihari's depiction, located in the divine mind.

When thinkers from our times develop these themes, they prefer to use an idiom based on infinite idealizations, possible minds and possible words, without reference to the divine. When linguists today try to learn specific points from their ancient Indian predecessors, they prescind from this divine-secular distinction. But the religious content of Bhartrihari regains importance when we look at the path that leads from his text to crucial features of mediaeval Indian history – a matter we touch upon briefly, later in this exposition.

Where can linguists who normally look to Panini's work for early Indian models learn anything specific in Bhartrihari? What is it about his model of language and knowledge that can help underwrite today's transition from formalist to substantivist versions of generative grammar?

The most direct answers to this question already offered in technical papers – Dasgupta (2003, 2005, 2006) – specify the conceptual and empirical challenge that unreformed formalists face within generative grammar. However, what may turn out to be the most crucial challenge has to do with the relation between Bhartrihari's *vaakya* "sentence" and his *pada* "syntactic word" in the languages known as polysynthetic, in which a single *pada* typically stretches to accommodate the content of an entire clause (for relevant discussion, see especially Dasgupta et al. 2000).

The fact that a polysynthetic word that looks more or less like *boy-girl-pumpkin-gave* can in some languages mean "The boy gave the pumpkin to the girl" is of intrinsic interest and has been studied a great deal. The transition from an unsupplemented Paninian or neo-Paninian formalism to a Bhartrihari-sensitive substantivism, however, lends some importance to the further fact that no known language, present or past, has ever allowed a *pada* to expand further and accommodate a bi-clausal structure. In other words, in no language do we see a word like *man-woman-told-boy-girl-pumpkin-gave*, which, had it been possible, would have meant "The man told the woman that the boy gave the pumpkin to the girl."

In other words, Bhartrihari's thesis about the *vaakya* being the natural unit of linguistic articulation – which is a matter of the *kaaraka* structure surrounding a verb – amounts to the claim that the clause is this natural unit. In contrast, a multi-clausal sentence is best conceptualized in terms of clauses counting as multiple attention domains bridged by trans-clausal mechanisms explicitly performing formal mediation between embedder and embedded clauses. On such a construal, a sentence, always potentially a multi-clausal entity, becomes a site of discourse in the sense of Dasgupta et al.'s (2000) articulation of substantivism.

In other words, a discourse-based theory that permits multiple exit points, and is thus not confined to a naive Paninian monolithic derivation, enables us to conceptualize inter-clausal relations in a complex sentence as comparable to interpersonal relations in a dialogue. Such a theory opens up the sentence and introduces sociality at the heart of one's theory of language and knowledge.

These issues continue to count as matters awaiting empirical clarification for formalist grammarians, who have yet to come up with a morphology-syntax interface theory that makes sense of the inability of polysynthetic words to include *kaaraka* grids surrounding two or more verb nuclei. In contrast, substantivism builds into its theoretical first principles the distinctive role of the single-verb *kaaraka* grid as the basis for *pada* to *pada* connecting operations, from which these properties of polysynthesis follow directly and without special stipulation.

The discussion so far has quite naturally focused on knowledge of language per se as we understand this domain today. In ancient India, this domain was not quite so sharply demarcated and flowed into questions of religion. It is really a formal point that, just as classical Indian grammar permitted only a disciplined, derivation-controlled form of exit from the linguistic into the semantic, and the substantivist supplement opens up the possibility of multiple exits, so also Bhartrihari's intervention opens up the religious validation of non-Vedic *prasthaanas*.

Space prohibits the proper pursuit of this theme in the present intervention. But a formally inclined reader will have noticed that the *miimaamsaka*'s account of the relation between the Vedic *karmakaanda* and its textual basis closely resembles – and is of a piece with – the grammatical knowledge system that the *Vedaangas* set up. Correspondingly, the substantivist opening up also has the impact of turning the Vedic text into a discourse and of switching on the multi-planar translation apparatus for the religious dimension as well. It virtually follows from such logic that the population will take over from the priests, and that a formation of the *bhakti* type, with its *pauraanika* adjuncts taking centre stage, will replace the formal Vedic orthodoxy with the postformal, translative multi-system of the Hinduism that comes to the fore in mediaeval India. Formally inclined readers can, without too much difficulty, read Bhartrihari's *Vaakyapadiiya* as a manifesto for that postformal order. Elaboration of this point will have to wait. As in many other domains, the clarification of the mediaeval presupposes further research on its ancient underpinnings.

Colophon

I would like to thank Sundar Sarukkai and the audiences at his 2006 and 2007 Philosophy Summer Courses at the National Institute for Advanced Study, Bangalore, as well as Nirmalangshu Mukherji and the audience at his 2007 Indian philosophy text reading exercise at the Department of Philosophy, University of Delhi, for useful feedback on pedagogic versions of these remarks. The usual disclaimers apply.

Notes

1 The transcription used in this chapter adheres to the following user-friendly conventions. In the case of words or names that have a standard or standardizable orthography in Roman script languages (such as the term Ayurveda), this orthography is used. For other expressions, there are two degrees of rigour. A word simply used in the English text appears with vowel length indicated by doubling the letter (as in the word *sarvanaama* for "pronoun") but with consonant articulation details such as retroflexion not shown. Real citations in Sanskrit use doubling for long vowels and tx dx nx sx for retroflex segments, rx lx for syllabic liquids, s' for the palatal sibilant, mx for the *anusvaara*, and hx for the *visarga*.

78

2 His words were: "Kim punar nityah s'abda a:hosvit ka:ryah? Samgraha etat pra:dha:nyena pari:ksxitam nityo va: sya:t ka:ryo veti. Tatrokta: dosxa:h prayojana:ny apy ukta:ni. Tatra tv esxa nirnayo yady eva nityo' tha:pi ka:rya ubhayatha:pi laksanam pravartyam iti" – cited by Subrahmanyam 1999, 21, also the source of our English rendering, apart from "originary" for Subrahmanyam's "eternal," and "derivative" for his "one to be derived."

3 His words were: "Ghatena ka:ryam karisyan kumbhaka:rakulam gatva:+ a:ha 'kuru ghatam ka:ryam anena karisya:mi' iti. Na tadvac chabda:n prayoksyama:nxo vaiya:karanakulam gatva:+ a:ha "kuru s'abda:n prayoksxy" iti. Ta:vaty eva+ artham artham upa:da:ya s'abda:n prayuñjate. Yadi tarhi loka esxu prama:nxam kim s'a:strenxa kriyate?" (Subrahmanyam 1999, 18).

4 "Raksxoha:gamalaghvasamxdeha:hx prayojanam" (Subrahmanyam 1999, 2).

5 For readers who may need to be keyed into the multi-planar organization Bhartrihari postulates, here are some rough glosses for the way he visualizes his planes of synthesis. *Vaikharii* is articulately segmented speech. *Madhyamaa* is linguistic form on a plane neutral between speaker and hearer and therefore prescinding from segmentation into sounds, words etc. *Pashyantii (pas'yantii)* is semantic form on a plane that renders the reality of the extralinguistic referent fully perceptible.

6 For relevant discussion, see Sarukkai (2005, 247–249).

Works Cited

Cardona, George. *Pa:nini: His Work and Its Traditions (Vol. I: Background and Introduction)*. New Delhi: Motilal Banarsidass Publishers Pvt Ltd., 1988.

Dasgupta, Probal. "Outgrowing Quine: Towards a Substantivist Theory of Translation." *International Journal of Translation*. Vol.1.2 (1989): 13–41.

———. "Antiopacity and Bangla Causatives: A Substantivist Approach." *The Yearbook of South Asian Languages and Linguistics*. Eds. Rajendra Singh et al. Berlin: Mouton de Gruyter, 2003. 47–70.

———. "Q-baa and Bangla Clause Structure." *The Yearbook of South Asian Languages and Linguistics*. Eds. Rajendra Singh et al. Berlin: Mouton de Gruyter, 2005. 45–81.

———. "Bikiron, Aakoronpokkho Aar Kaayaapokkho" [In Bangla: "Lexical Scope, Formalism and Substantivism"]. *Saahitto Porisat Potrikaa*. Vol.113.1–2 (2006): 147–157.

Dasgupta, Probal, Alan Ford, and Rajendra Singh. *After Etymology: Towards a Substantivist Linguistics*. Munich: Lincom Europa, 2000.

Joshi, S.D., and Saroja Bhate.*The Fundamentals of Anuvrtti*. Pune: University of Pune, 1984.

Joshi, S.D., and S.D. Laddu. (eds.). *Proceedings of the International Seminar on Studies in the Asta:dhya:yi: of Pa:nini (Held in July 1981)*. Pune: University of Pune, 1983.

Katre, S.M. *Asta:dhya:yi: of Pa:nini (Roman Transliteration and English Translation)*. New Delhi: Motilal Banarsidass Publishers Pvt Ltd., 1989.

Pillai, K. Raghavan. *The Vakyapadiya: Critical Texts of Cantos I and II (with English Translation, Summary of Ideas and Notes)*. New Delhi: Motilal Banarsidass Publishers Pvt Ltd., 1971.

Sarukkai, Sundar. *Indian Philosophy and Philosophy of Science*. New Delhi: Project of History of Indian Science, Philosophy and Culture, 2005.

Subrahmanyam, P.S. *Pa:ninian Linguistics*. Tokyo: Institute for the Study of Languages and Cultures of Asia and Africa, 1999.

Subramania Iyer, K.A. *Bhartrhari: A Study of the Va:kyapadi:ya in the Light of Ancient Commentaries*. Pune: Deccan College, 1969.

7

INDIAN DEDUCTIVE SYSTEM

The logical basis of Indian sciences

Nirmalya Guha

The Indian deductive system consists mainly of three aspects: inference (*anumiti*), the cognitive validator (*tarka*), and pure logic (*yukti*). The whole system is built on the idea, which may be called blocking (*pratibandhana*). A piece of cognition C blocks another, say D, when C does not allow D to be generated. The general rule is: The cognition of *x* and the cognition of the absence of *x* mutually block each other (unless either of them is stronger than the other).[1] Thus the cognition, 'there is no water in the pot' blocks the genesis of the cognition, 'there is water in the pot.'

In the first half of the chapter, I shall briefly discuss inference and the cognitive validator. *Yukti* will be discussed in the sub-section on the application of Indian deductive rules in mathematical sciences. However, the focus of the first half of the chapter will be the common database and data management. In the second half, I will try to show how some of the Indian sciences have used the deductive system.

The Indian deductive system

The principal unit of the Indian deductive system – actually of the entire Indian epistemology – is cognition (*jñāna*). A **cognition** happens to me.[2] Thus it is not my imagination or thought, which I can control. Upon seeing a car, I have the cognition 'this is a car.' Upon seeing a coil-like thing in a dark room, I have the doubt-cognition 'is this a snake or a rope?' Upon mistaking a rope for a snake, I have the illusory cognition 'this is a snake.' But I have no control over their content, occurrence or non-occurrence. I cannot make a cognition happen or prevent it from happening. I cannot even change its content. The **content of a cognition** about a car is the car, captured by the cognition.

Some cognitions are valid, and some are not. What is a valid cognition?

D1: The **cognition**, '*x* is *y*' is **valid** *iff* *x* really is *y* (i.e., *x*'s being *y* is a fact).[3]
Example: The cognition 'this is a snake' is valid when the object seen by the subject is really a snake.

According to the school of Indian logic (Nyāya), the factors that can generate valid cognitions are of four types: the sense organs, inference, testimony, and analogy. The study of the validity-makers is Indian epistemology (*pramāṇa-śāstra*).

There is a significant difference between inference and the cognitive validator, although both of them are cognitive elements. The former generates epistemic information, whereas the latter evaluates and licenses informative contents. I have observed that every renate is a cordate.[4] Now just by seeing that somebody is a renate I inferentially know that they are also a cordate. Their being a cordate is new information generated by inference. But the cognitive validator works differently. Ockham's razor is such a tool and is known as economy (*lāghava-tarka*) in the Indian context. It compares different theories/options/possibilities and chooses one over the others.

Inference (anumiti)

The central idea of the Indian inference is pervasion (*vyāpti*). Here is a rough definition of pervasion:

D2: **y** **pervades** x (or x is pervaded by y) *iff* all cases of x are cases of y. Example: Every case of having a kidney is a case of having a heart.

What I infer is the **Target** (*sādhya*), and the locus, in which I infer the Target, is the **Site** (*pakṣa*) of my inference. The object, from which I infer the Target is the **Reason** (*hetu*). In the inference, 'Tom has a heart, since he has a kidney,' the Site (S), Target (T), and Reason (R) are Tom, the heart, and the kidney, respectively. Without knowing the pervasion between the heart and the kidney, I cannot infer the former from the latter. Thus knowledge of pervasion is a necessary condition for inference. There is another condition:

D3: The Reason has the property called **site-location** (*pakṣa-dharmatā*) *iff* the Reason is located in the Site. Example: 'Tom has a kidney' represents the site-location of the kidney in the Site (i.e., Tom).

Thus the inferential assertion 'Tom has a heart, since he has a kidney' also asserts pervasion (i.e., every case of having a kidney is a case of having a heart) and site-location (i.e., Tom has a kidney). Every inference is just an instantiation of the basic schema, 'S has T since S has R' or 'S is a T-possessor, since S is an R-possessor.' The stock example the Indian logician uses is 'the hill has fire since it has smoke,' which is derived from the schema on replacing 'S,' 'T,' and 'R' by 'the hill,' 'fire,' and 'smoke,' respectively. Thus this specific inference is an *instance* of the basic schema. Every inference is valid by default; it is defective – and hence blocked – if it has a defect.

Inferential defect (hetvābhāsa)

Here is a rough definition of the inferential defect:

D4: The content of the cognition C is the (**inferential**) **defect** of the inference F *iff* C contradicts any assertion made by F.[5] Example: The cognition 'there is no fire on this hill' contradicts the main assertion 'this hill has fire' made by the inference 'this hill has fire, since it has smoke.' Therefore, the content of the cognition is the defect of the inference. The content is the absence of fire from that very hill.

Soon we shall see that D4 is not strong enough to define the inferential defect. The defect should be a real one. Suppose Tom has all the symptoms of jaundice. But Harry, who does not have any knowledge of medical science, says that Tom does not have jaundice. The physicians, who have seen Tom, would definitely say, "No! Harry is wrong. Tom surely has jaundice since he has all the symptoms." In this case, despite contradicting the inference of jaundice, the imaginary absence of jaundice does not make the inference faulty. Hence it, i.e., the absence, is not an inferential defect. In order to get around these cases, we must modify the definition in the following way:

D5: The content of the valid cognition C is the (**inferential**) **defect** of the inference F *iff* C contradicts any assertion made by F.[6] Example: The valid cognition 'there is no smoke on this hill' contradicts the site-location of the inference 'this hill has fire, since it has smoke.' Therefore, the content of the valid cognition, i.e., the absence of smoke from that very hill, is the inferential defect in this case.

Harry did not have a valid cognition about Tom's jaundice. He – Harry – just thought Tom did not have it without even having the right inquiry. Since the absence of jaundice was not the content of a valid cognition, it could not block the perfect inference about Tom's jaundice and, hence, was not an inferential defect.

There are inferential defects of several types.

(1) **Deviation** (*vyabhicāra*) is a defect that challenges the pervasion of an inference. Suppose one infers, 'that hill is smoky, since that is fiery.' The pervasion-claim of this inference is: anything that has fire has smoke. However, this is wrong, since a red-hot iron ball (which is fiery without being smoky) is evidence against this. Here the Reason is disloyal to the Target (i.e., there is at least one case in which the Reason 'fire' is not accompanied by the Target 'smoke'). In order to be pervaded by the Target, the Reason must always be accompanied by the Target. Deviation is a disloyal Reason unaccompanied by the Target.

(2) The **unestablished Reason** (*svarūpāsiddhi*) is a defect that targets the site-location. The real absence of smoke from the hill is the unestablished Reason for the inference 'the hill has fire, since it has smoke.'

(3) **Opposition** (*bādha*) is a defect that targets the principal assertion of the inference, i.e., '*S* has *T*.' Suppose Tom claims that 'this hill has fire, since it has smoke' and Harry correctly states that 'I have seen the entire hill. There is no fire anywhere.' Here, the hill without fire is the opposition.

(4) The **unestablished Site** (*āśrayāsiddhi*) is a defect that partially targets the principal inferential assertion by targeting a property of the Site. The unestablished Site of the inference 'the golden hill has fire' is the absence of golden-ness from the hill.[7]

It is easy to understand that 'the content of a valid cognition' is nothing but a fact. A doubt or illusion is invalid, since it does not have any factual content (i.e., nothing in reality corresponds to the content of such a cognition). The classical Indian worldview is cognition-centric. An object here is seen as 'the referent of a word' (*padārtha*). In this framework, when I claim that there is a 'fact', my claim must be based on some knowledge, some cognition that is valid. Thus a 'fact' becomes merely the content of a valid cognition. This "cognition-first" orientation is a unique feature of the Indian philosophical thinking. In the next subsection, we shall see that the definition of the inferential defect cannot be expressed in terms of facts. But the revision of the definition is also problematic.

Data management: the common database

Guha (2015, 569) describes an imaginary medical condition called *Coccinistercus* and the following (imaginary) case. Some physicians have observed that somebody who eats strawberries every day suffers from *Coccinistercus*, and somebody who eats blackberries every day never suffers from this disease. Both the observations are un-countered. And they seem to be innocent. The problem begins when the physicians find a little boy who eats both strawberries and blackberries every day. Does he have *Coccinistercus*? The physicians now face a dilemma since the following inferences counterbalance each other: 'the boy has *Coccinistercus*, since he eats strawberries every day' and 'the boy does not have *Coccinistercus*, since he eats blackberries every day.' Both the inferences cannot be true. This means that both the inferences cannot have factual support. In that case, even D5 is not strong enough as a definition of the inferential defect. Why? According to D5, the defect of the inference 'the boy has *Coccinistercus*' is the *fact* that he eats blackberries every day, and regular consumption of blackberries implies the absence of *Coccinistercus*. But the pervasion 'somebody who eats blackberries every day never suffers from *Coccinistercus*' may or may not be factual. The point is that both the rivaling pervasions seem to be correct, since neither of those has a counter-example. But they both cannot be correct. Above all, both the inferences are defective and do block each other. Such a defect is called **counter-balance** (*satpratipakṣa*), and it is caused by the mutual blocking of two equally strong (*tulya-bala*) contradictory inferences.

Since the counter-balance as an inferential defect is not factual, we must modify the general definition of the inferential defect.

D6: The content of the un-invalidated cognition C is the **(inferential) defect** of the inference F *iff* C contradicts any assertion made by F.

D7: The **cognition** C is **un-invalidated** *iff* the invalidity of C has not been confirmed (*agṛhīta-aprāmāṇyaka*[8]). Example: In the case of *Coccinistercus*, both the cognitions of pervasion are un-invalidated, since nobody could confirm that they were invalid.

We must note that every valid cognition is also un-invalidated, although every un-invalidated cognition is not valid. Now the question is: Is this definition unproblematic? Perhaps it is not![9]

Had counter-balance not been there, D5 could be an acceptable definition of the inferential defect. But counter-balance is very important, since our knowledge-world is not fully consistent. Since, in the above case, each of the pervasions works fine in isolation, we should not and cannot banish those pervasions from our database. Still there is a conflict when they come together. The Indian deductive system wants to manage the (partially) inconsistent data in a clever way. That is why it has the counter-balance. Suppose *Coccinistercus* turns out to be a real disease, and also suppose that there is a real case of daily consumption of both strawberries and blackberries. The physicians would not be able to pronounce any verdict in this case. How to define this near-real situation? Now the counter-balance becomes useful.[10] Nevertheless, the replacement of "valid" by "un-invalidated" still seems to be problematic.

Let us consider the following case. Tom infers, 'this hill has fire, since it has smoke.' But Harry does not know that the aforementioned hill has smoke, even though it really has smoke. Naturally Harry tells Tom, "Your wrong inference has the unestablished Reason, since the Reason is not there in the Site." It is true that according to Harry's database Tom's inference is defective. It is because smoke's absence from the hill is the content of a cognition (i.e., 'the hill does not have smoke') whose invalidity is not known to Harry. Hence the cognition is surely un-invalidated for him. Yet we intuitively understand that Tom is right here.

I think we can save both the term "un-invalidated" and the intuition that Tom's inference is defectless by interpreting the former slightly differently.

Let there be a commonly accessible database for both Tom and Harry. And let us redefine "un-invalidated" in the following way:

D8: The **cognition** C is **un-invalidated** *iff* the invalidity of C has not been confirmed with respect to the common database accessible to all the parties of the debate. Example: Both the cognitions 'somebody who eats strawberries every day does suffer from *Coccinistercus*' and 'somebody who eats blackberries every day never suffers from *Coccinistercus*' are un-invalidated, since on the common database accessible to all physicians their invalidity has not been confirmed.

This new interpretation of un-invalidation does not allow the absence of smoke from the hill of Tom and Harry to infect Tom's inference 'this hill has fire, since it has smoke.' The hill is accessible to Harry. If he checked the hill properly, he would have seen smoke there, provided his senses were working fine. He did not check the hill properly. It is his personal failure and not a failure of the commonly accessible database.

Cognitive validator

I have already mentioned that the cognitive validator does not generate any epistemic content; it just evaluates and validates a content. Since they do not generate any information, they are not part of the Indian epistemology in a strict sense. I would call them supra-epistemic. Although they do not play an epistemic role, they sit on top of the epistemic system, and when there is informative conflict, they manage the data. Thus the datum that comes out of the deductive system does not disturb our knowledge-world.

1   *Reductio-ad-absurdum* (*prasaṅga*): We know that by adding 1 to a number we get a larger number. Let us suppose that the largest number exists. Let that number be N. Since N is a number, (N + 1) must also be a number. But our initial supposition was that no number is larger than N. Thus by assuming that the largest number exists, we are led to an absurd position that is marked by a contradiction (i.e., no number is larger than N and (N + 1) is larger than N). An assumption that leads one to contradiction must be wrong. Hence, its denial must be right. Therefore, we conclude that there is no largest number. The cognitive validator that rejects the absurd assumption and establishes the denial of the rejected assumption is the *prasaṅga tarka*.

2   **Infinite Regress** (*anavasthā*): The theory 'In order to apprehend a cognition, one needs another cognition' is to be rejected by the *anavasthā tarka*. For this theory will not allow one to rest anywhere and apprehend a specific cognition. In order to apprehend the cognition C, one needs

D, and in order to apprehend *D*, one needs *E* and *ad infinitum*. Thus this *tarka* rejects theories that involves the undesired non-stoppage.

3 **Economy and Profligacy** (*lāghava-gaurava*): Let us compare the following theses: (a) 'all young humans are mortal, and all middle aged humans are mortal, and all old humans are mortal,' and (b) 'all humans are mortal.' Both (a) and (b) seem to be true. Still the *lāghava-gaurava tarka* compares the two theses and prefers (b) over (a) since (b) is more economical than (a).

4 **Default Value and Exception** (*utsarga-apavāda*): I know that Tom is trustworthy and his senses are working fine. That being the case, I will accept whatever Tom states by default. I will not accept his statement when there is a stronger opposition. Suppose he tells me that there is no Taj Mahal in Agra, and I go and see it there. Here my perceptual evidence against Tom's testimony is the *apavāda*. The *utsarga tarka* makes me accept Tom's statements, whereas the *apavāda tarka* tells me when I should not do so.

Our database is not all perfect. It may have inconsistencies. Sometimes physical checking of data is not possible. *Tarka*s help us make decisions when there is data-conflict or impossibility of physical checking. I do not wish to say that whenever there is inconsistency or the absence of checkability in our database, a *tarka* helps us arrive at the right conclusion. The *tarka* operates only when the conflict of information or impossibility of checking follows a certain pattern. For example, it is not possible for us to physically check whether the largest prime number can exist by enumeration. But the *prasaṅga tarka* helps us do that by demonstrating that the existence of the largest prime number would somehow invoke contradiction while the human mind wants to avoid contradiction. Guha (2012) claims that the human mind has an *a priori* list of cognitively undesired elements. Inconsistency, profligacy, non-acceptance of the default value in absence of counter-evidence, non-stoppage of a process, etc. are on the list. On the contrary, consistency, economy, acceptance of the default value in absence of counter-evidence, stoppage, etc. constitute the *a priori* list of cognitively desired elements. When a piece of information or theory contains some *a priori* undesired elements, the mind invokes a *tarka* to resolve the conflict.

The Indian deductive system: an overview

In this subsection, I shall try to give a holistic picture of the Indian theory of knowledge. I shall have to use contemporary terms for drawing the picture. Classical Indian philosophical ideas are profound and meticulous. Still, I feel that Indian thinkers were not very interested in giving an overview of their thoughts that were at the same time widespread and networked. But to the modern mind, it is important to have an overview of every system.

According to the Indian School of Logic (Nyāya), the epistemic subject (who owns ideas, attains knowledge, mistakes something for something else, or doubts something) is mainly the soul (*ātman*) assisted by the mind (*manaḥ*), sense organs (*indriyas*), and a few cognitive abilities.[11] The sense organs are their windows through which raw data enter the system. These are mainly perceptual data. The epistemic tools such as inference (*anumāna*), testimony (*śabda*), and analogical reasoning (*upamāna*) work on the perceptual data and generate mediate cognitions. Suppose the subject has observed that every case of smoke is a case of fire and now sees smoke on a hill; through inference they will combine these two pieces of data and inferentially attain the cognition 'this hill has fire.' If both the pieces of foundational data are correct, the subject would have a valid inference. When the subject receives raw data only about the common property of two objects *x* and *y*, they have the doubt, 'is it *x* or *y*?' Sometimes the subject receives partial data from the object and fills up the gap by supplying it with something from memory. These are the cases of illusory cognitions. Thus one sees a coil-like thing and cognizes, 'this is. . . .' Now one's mind would fill up the gap with the memory of snake-ness (*sarpatva*) due to the resemblance between the object and a snake, and one would illusorily cognize, 'this is a snake.'

The deductive system, which is a subsystem of the cognitive system, can be seen as a data processor. It receives raw data and generates inferences and then runs those through the inferential filters. These filters are the blockers, which, in this inferential context, are the inferential defects. The filters too are part of the existing database. A piece of inference gets validated when it passes through all the filters unblocked. It then gets internalized by the database. Suppose one who knows that it is not true that every case of fire is a case of smoke infers, 'this hill has smoke since it has fire.' The inference along with all its assertions (such as the main assertion 'this hill has smoke,' the pervasion 'every case of fire is a case of smoke' and the site-location 'this hill has fire') gets verified against the database of the subject. In this case, the pervasion is clearly contradicted by the database that contains a red-hot iron ball, which is fiery without being smoky. Thus the inference gets blocked. We may note here that the filter itself is a datum. When one infers, 'Tom is a cordate since he is a renate,' nothing blocks one's inference, and hence it enters one's database. This is how the database gets modified.

The supra-epistemic aspect or *tarka* works in a slightly different way. It operates on raw data generated by the epistemic tools or the mind. It compares and evaluates data and finally licenses one. For example, perception does not decide between the following options: (a) 'all young humans are mortal, and all middle-aged humans are mortal, and all old humans are mortal' and (b) 'all humans are mortal.' It is the economy *tarka* that prefers

(b) over (a). A *tarka* can work on mental data too. Suppose one comes up with the thesis 'There is no largest prime number.' Then one wants to know whether one's thesis is correct. The *prasaṅga tarka* helps one see that this thesis is correct by discovering a contradiction in the womb of the argument that begins by assuming that there exists the largest prime number. Once a *tarka* supports a datum, mental or epistemic, it enters the database. Thus a lot of raw data gets rejected either by the inferential blocker or by the supra-epistemic blocker. The database projects the knowledge-world to which the subject has direct access. What we consider our objective world is nothing but this knowledge-world.

The application of the deductive rules

The argumentative style of almost every Indian academic discipline is based on the rules of the Indian deductive system. I will discuss the application of the deductive rules by two Indian sciences, namely the medical and mathematical sciences.

Medical science: carakasaṃhitā

The eighth chapter of *Vimānasthāna*, i.e., the third section of *Carakasaṃhitā* (CS), briefly discusses the deductive system. The author of CS enlists forty-four elements of debate the physician must learn. The assertion of one's position (*pratijñā*), logical demonstration of the assertion (*sthāpanā*), logical demonstration of the opposition (*pratiṣṭhāpanā*), Reason (*hetu*), example (*dṛṣṭānta*), general rule (*upanaya*), conclusion (*nigamana*), refutation of an argument by demonstrating similarity or dissimilarity (*uttara*), provisional or final hypothesis (*siddhānta*), inference (*anumāna*), verbal inconsistency (*savyabhicāra*), inferential defect (*ahetu*), etc. are among those forty-four elements of debate.

The author of CS says:

> Inference is reasoning (*tarka*) based on logic (*yukti*). The stomach-fire is inferred from the digestive power, the physical strength from the performance of physical exercise, and the proper functioning of the sense organs from the fact that one hears and sees things properly.[12]

The stomach-fire etc. cannot be perceived. Still the physician would have to know whether those are working fine in the body of their patient. Here inference is the only tool.

In CS, the inferential defects are presented slightly differently.[13] There are three defects, namely *prakaraṇasama*, *saṃśayasama* and *varṇyasama*.

1   *Prakaraṇasama* is nothing but the counter-balance.
2   The example of *saṃśayasama* cited in CS is the following:

> "This man knows only a part of the medical science. Due to his partial knowledge of the science, somebody doubts, 'Is this man a physician or not?' Somebody else then asserts, 'This man must be a physician since he knows a part of the medical science.' This inference does not present a Reason that would dispel the doubt. Here the fact that causes the doubt is used as the Reason. Such an inference suffers from the defect called *saṃśayasama*."[14] Here the problem is that there is no established rule that states: 'whoever knows a part of the medical science is a physician.'

That being the case, the alleged Reason is not supposed to work.

3   When the example (*dṛṣṭānta*) of an inference does not present an established Reason (*varṇyāviśiṣṭa*[15]), the inference suffers from the defect called *varṇyasama*. When one argues, 'This hill has fire, since it has smoke, like my kitchen,' one's example is well-established. In a kitchen, fire accompanies smoke. But if somebody infers, 'Tom has jaundice, since his hands are weak, like Harry,' their inference gets flawed due to the fact that there is no confirmation that Harry's hands were weak when he had jaundice. It is a failure of the example.

Mathematical science

In this section, I would like to introduce another aspect of the Indian deductive system, *yukti* or *upapatti*. I wish to translate this as "pure logic" since it is not based on empirical observation. In a broad sense, this is also inference, since it helps one conclude y from x when one knows that every case of x is a case of y. However, there is a significant difference between *anumiti* and *yukti*: *anumiti* is based on observation, whereas *yukti* is not. Narasimha (2007, 528) shows that *yukti* is absolutely central to the methodology of every Indian mathematician. In *Siddhāntadarpaṇam*, while giving a proof for Pythagoras' theorem, Nīlakaṇṭha says: 'All this [understanding] is from *yukti*, not from scriptures. Thus the intelligent can see this without depending on the activities of their eyes, hands, etc., just by using their mind.'[16] I quote Srinivas' (2005, 218–219) translation of a passage from *Bījapallava*, Kṛṣṇa Sarvajña's commentary on *Bījagaṇita* of Bhāskara:

> How can we state without proof (*upapatti*) that twice the product of two quantities when added or subtracted from the sum of their squares is equal to the square of the sum or difference of those quantities? That it is seen to be so in a few instances is indeed of no

90

consequence. Otherwise, even the statement that four times the product of two quantities is equal to the square of their sum, would have to be accepted as valid. For, that is also seen to be true in some cases. For instance, take the numbers 2, 2. Their product is 4, four times which will be 16, which is also the square of their sum 4. Or take the numbers 3, 3. Four times their product is 36, which is also the square of their sum 6. Or take the numbers 4, 4. Their product is 16, which when multiplied by four gives 64, which is also the square of their sum 8. Hence, the fact that a result is seen to be true in some cases is of no consequence, as it is possible that one would come across contrary instances (*vyabhicāra*) also. Hence it is necessary that one would have to provide a proof (*yukti*) for the rule that twice the product of two quantities when added or subtracted from the sum of their squares results in the square of the sum or difference of those quantities.

The *upapatti* is demonstrated only through *yukti*, which mainly looks for some general pervasion-like laws that are not based on empirical data. But, like pervasion, these laws too are subject to blocking. The pseudo-law 'Four times the product of two quantities is equal to the square of their sum' is wrong since it has deviation. The deviation can be seen in the cases that involve the pairs of unequal numbers such as 2, 3 or 3, 4.

The mathematicians used *tarkas* also. Once again I quote Srinivas' (2005, 230) translation of a passage from *Bījapallava*:

> Let it be stated by you who claim that a negative number is a square as to whose square it is: Surely not of a positive number, for the square of a positive number is always positive by the rule. . . . Not also of a negative number. Because then also the square will be positive by the rule. . . . This being the case, we do not see any such number whose square becomes negative.

Through this *prasaṅga tarka* Sarvajña proves that a negative number cannot be a square. It seems that inference based on observation or *anumiti* is not very useful in mathematical sciences, although it is extremely useful for astronomy.

Almost all the academic disciplines in classical India used the rules of the Indian deductive system. As the system evolved, the disciplines kept updating their polemical styles. After the emergence of Neo Logic (Navya-nyāya), most of the disciplines adopted the complicated style of the Neo Logician. Thus, the argumentative style of an Indian academic discipline developed systematically and gradually. Unfortunately, the colonial period witnessed the end of the intellectual growth of classical India.

Unlike the Indian system, the Western logical system is axiomatic. It consists of a few foundational self-evident truths (i.e., axioms) and some rules. A theorem can be cooked by using those axioms in a rule-governed way. Something that cannot be cooked this way is not a theorem. On the contrary, the Indian system allows one to cook anything. It will be validated if it is not blocked by a valid rival. Thus, blockers were the main players in the Indian system. Somehow, colonial India preferred the axioms to the blockers.

Notes

Acknowledgement: I am deeply indebted to Dr. Vinita Chandra and Dr. Andrey Klebanov for their corrections and comments on the first draft of this chapter.

1 Jhalakīkar (1928, 433): *tad-vattā-buddhim prati tad-abhāva-vattā-niścayatvena . . . pratibandhakatvam.*

Translation: The doubt-free cognition 'there is no x' blocks the cognition 'there is x.'

2 For the sake of convenience, I shall treat 'cognition' as a count noun in this chapter. The technical terms in bold type imply that they are being defined or contextualized there.

3 '*Iff*' is the standard abbreviated form for 'if and only if.'

4 A renate is a creature with a kidney, and a cordate is a creature with a heart.

5 *Siddhāntamuktāvalī* (SM), The Chapter on Inference (Śāstrī 2015, 238): *yad-viṣayakatvena jñānasya anumiti-virodhitvaṃ tattvam.*

Translation: The property of being an inferential defect is the property of being the content of a cognition that opposes an inference.

6 *Siddhāntamuktāvalī* (SM) (p. 239): *yad-viṣayakatvaṃ ca yādṛśa-viśiṣṭa-viṣayakatvam.*

Translation: 'The property of being the content' means 'the property of being the qualified content.'

Comment: The cognition 'this is a snake' has three constituents: the referent of the word 'this' as the qualificandum (*viśeṣya*), snake-ness as the qualifier (*viśeṣaṇa*), and inherence as the cognitive relation (*saṃsarga*). The content of a cognition is qualified *iff* the qualifier of the cognition really qualifies the qualificandum. In the case of the aforementioned cognition, its content is qualified *iff* the object that appears as 'this' in the cognition really possesses the qualifier-property snake-ness. If the cognition is illusory, then 'this' does not really have snake-ness. Maybe 'this' is actually a snake. something else. iff ioned cg, sciences used the Indian deducitve and data-management. onsisting of ir. kberry. ions a a rope (i.e., something, which has rope-ness). snake. something else. iff ioned cg, sciences used the Indian deducitve and data-management. onsisting of ir. kberry. ions

7 Similarly, 'the hill has golden fire' has the unestablished Target (*sādhyāsiddhi*) because golden-ness is always absent from fire.

8 The author of SM uses this Sanskrit phrase in the chapter on inference. See Śāstrī (2015, 247).

9 As far as I know, this problem has not been critically discussed in the published literature.

10 Once a neuroscientist, whose name I unfortunately forgot, attended my talk on the inferential defect and told me something very interesting. She suggested that perhaps the schizophrenic patient finds it difficult to make a decision because almost every piece of their knowledge K gets counter-balanced by an illusory cognition that opposes K.

11 The epistemic subjects of some Indian philosophical schools are constitutionally different from the Nyāya subject. But their deductive systems work in more or less the same way as the Nyāya system works.
12 Pāndeya (1969, 641): *anumānaṃ nāma tarko yuktyapekṣaḥ. yathā agniṃ jaraṇaśaktyā, balam vyayāmaśaktyā, śrotrādīni śabdādigrahaṇena ity evam ādi.*
13 Ibid. (p. 646): *ahetur nāma prakaraṇasamaḥ, saṃśayasamaḥ, varṇyasamaś ceti.*
14 Ibid. (p. 647): *ayam āyurvedaikadeśam āha, kiṃ tv ayaṃ cikitsakaḥ syān na veti saṃśaye paro brūyāt – yasmād ayam āyurvedaikadeśam āha, tasmād cikitsako 'yam iti. na ca saṃśayacchedahetuṃ viśeṣayati. eṣa cāhetuḥ. na hi ya eva saṃśayahetuḥ sa eva saṃśayacchedahetur bhavati.*
15 Ibid.
16 For the Sanskrit text, see Sarma (1976, 24).

References

Guha, N. "Tarka as Cognitive Validator." *Journal of Indian Philosophy*. Vol.40.1 (2012): 44–66.

———. "A Monstrous Inference Called Mahāvidyānumāna and Cantor's Diagonal Argument." *Journal of Indian Philosophy*. Vol.44 (2015): 557–579.

Jhalakīkar, B. *Nyāyakoṣa or the Dictionary of Technical Terms of Indian Philosophy*. Pune: The Bhandarkar Oriental Research Institute, 1928.

Narasimha, R. "Epistemology and Language in Indian Astronomy and Mathematics." *Journal of Indian Philosophy*. Vol.35 (2007): 521–541.

Pāndeya, G. (ed.). *The Caraka Saṁhitā of Agniveśa: Revised by Caraka and Dṛḍhabala*. Varanasi: The Chowkhamba Sanskrit Series Office, 1969.

Sarma, K. V. (ed.). *Siddhāntadarpaṇam of Nīlakaṇṭhasomayājī with Auto-Commentary*. Hoshiarpur: Panjab University, 1976.

Śāstrī, H.S. (ed.). *Nyāyasiddhāntamuktāvalī of Viśvanātha Pañcānan Bhaṭṭāchārya* (Reprint). Varanasi: Chaukhamba Prakashan, 2015.

Srinivas, M.D. *Proofs in Indian Mathematics*, 2005. http://iks.iitgn.ac.in/wp-content/uploads/2016/02/Proofs-in-Indian-Mathematics-MD-Srinivas-2005.pdf.

8

MOTION INTERPRETED

A bridge between science and spirituality

Madhumita Chattopadhyay

A journey from science to spirituality at first sight seems to be an impossible one since it is believed that they belong to two completely different domains. "Science," as has been defined in the *Chambers 20th Century Dictionary*, is "knowledge ascertained by observation and experiment, critically tested, systematised and brought under general principles." In the *Concise Oxford Dictionary*, science is described as a branch of knowledge, especially one that can be conducted on scientific principles; it is an organized body of the knowledge that has been accumulated on a subject. What we can gather from these definitions is that science is a systematic rational inquiry on a subject and its method is one of observation and experiment. On the basis of the data obtained through observation and experimentation, it tries to arrive at some general conclusions which are equally valid for all normally constituted minds placed in that situation. Science, mainly, is concerned with the physical aspect of nature. It is widely believed that science is objective and scientific knowledge is reliable and is established as true. Personal opinions and idiosyncratic speculations have no role in it. As distinguished from science, spiritualism is believed to deal with something non-physical. It implies the mind-body dichotomy, indicating a separation between the body and the soul.

The term "spirituality" refers to the doctrine that nothing is real except the soul or spirit, and such soul or spirit is beyond the range of ordinary observation or any scientific experiment. Science takes as its basis empirical, repeatable observations of the natural world and accordingly puts aside the idea that relies on supernatural forces as beyond the scope of science. As such modern science stands in sharp contrast to spirituality. The latter is considered as the anti-thesis of science and is equated with mysticism, which is characterized as irrational and anti-scientific. But this is not the only sense in which the term "spirituality" is to be understood. "Spirituality" is a word which is used in a variety of contexts to mean different things for different people at different times and in different cultural backgrounds. For instance, in different contexts the term "spirituality" may mean any of the following: (1) a quest for wholeness, (2) a belief in a higher being or beings, (3) a belief in some level of transcendence, and (4) a belief that there is more to life than the material or practical and so on. In the context of Buddhist philosophy the term "spirituality" is used more or less in the third sense to denote a stage of reality which cannot be described in empirical terms, a stage which ordinary language fails to reveal. This feature of indescribability is the characteristic feature of this spirituality and hence is distinguished from the stage of science, which always attempts to establish truths in terms of concepts and theories.

Though in Buddhist philosophy, especially in the system of Nāgārjuna, there seems to be an opposition between the level of empirical knowledge, which may be described to constitute the world of science, and the level of spirituality which transcends all empirical knowledge, the two are not considered to be poles apart. Rather it is shown that a proper understanding of the level of science leads us to the level of spirituality. This transition becomes very much obvious if we concentrate on their analysis of motion.

Motion is one of the most important concepts in physics. It is dealt with in the area called dynamics. Nāgārjuna, in the second chapter of his *Mūlamadhyamakakārikā*, analyzes the concept of motion from the standpoint of the moving agent, the act of moving, and the place being moved upon and points out that from none of them can we consistently speak of motion. It may be noted here by way of reference that philosophers from the early days have tried to show that this concept of high-energy physics, which assumes a line to consist of an infinite number of extensionless points, when expressed in ordinary terms is likely to yield fictions or paradoxes. The pioneering work in this area was made by the Greek philosopher Zeno of Elea. Zeno of Elea tried to show the impossibility of motion by drawing our attention to certain important paradoxes related to motion. In India, the Buddhist philosopher Nāgārjuna, with the help of his dialectic method, has shown the meaninglessness of the concept of motion and other concepts that are associated with motion. A trend has been noticed nowadays among modern Indologists, especially of the West, to compare Nāgārjuna's arguments with those of Zeno regarding motion. How far such comparisons are justified or not is not the concern of this chapter. The objective of this chapter is a bit different. It is to show that Nāgārjuna's approach here is not simply to analyze the concept of motion to point out its paradoxical character but at the same time to show how this concept can serve as a bridge to lead one from the world of facts/world of objects to a world beyond/a transcendental world which may be described as the world of spirituality.

The present chapter is divided into three sections. In the first one, I will discuss in brief some of the paradoxes regarding motion stated by Zeno. Though this section is not completely related to our present subject, it will help us to understand how philosophers have pointed out the paradoxical consequences of such an important scientific concept through rational analysis. In the background of this analysis it will be easier for us to understand Nāgārjuna's treatment on the subject. The second section will deal with Nāgārjuna's own analysis of motion as found in the chapter entitled "*Gatāgataparīkṣā*." Going through such analysis, we will be able to point out the paradoxes which Nāgārjuna has pointed out against motion. In the third section we will show how Nāgārjuna's analysis of motion ultimately transcends the scientific realm and leads to the world of spirituality.

Zeno's paradoxes

The Greek philosopher Zeno of Elea was well known for his paradoxes. Even Aristotle honoured him by regarding him as the "founder of dialectics." Of the several paradoxes formulated by Zeno, those concerning motion have a different status. Zeno's aim has been to re-describe motion in such a way that it can be shown to be impossible. Aristotle in his *Physics* mentions four forms of Zeno's argument regarding motion. Among them Dichotomy asserts the non-existence of motion on the ground that that which is in locomotion must arrive at the halfway stage before it reaches the goal. That is, it is impossible to complete an arbitrary journey from A to B, to start at A, move to B and then stop, for:

(P_i) A journey from A to B is a series of sub-journeys with no last member i.e., from A to ½ AB, from ½ AB to ¾ AB . . .

And

(P_{ii}) It is impossible to complete a series of sub-journeys with no last members.

Hence, the journey from A to B is never completed.

The second argument concerning motion is the so-called Achilles, and it amounts to this: in a race the quickest runner can never overtake the slowest, since the pursuer must first reach the point where the pursued started, so that the slower must always hold the lead. Suppose Achilles is running faster than the tortoise. Then as the race proceeds, the distance between Achilles and the tortoise decreases at a rate determined by the relative velocities. Let the distance between them initially be AB, and suppose their relative velocities are such that after ½ minute it has diminished to ½ AB, after ¼ minutes to ¼ AB, . . . and so on. Achilles" task then will be to complete this infinite series of lead narrowings ½ AB, ¼ AB, . . .

The third argument concerning motion is the one which shows that when an animal in locomotion cannot move, it is at rest. The argument then proceeds as follows:

1 Everything always rests when it occupies a space equal to itself.
2 What is moving is always in the present.
3 What is in the now occupies a space equal to itself.
4 What is moving always occupies a space equal to itself.
Therefore,
5 What is moving is always at rest (from 4 & 1)
6 Time is composed of indivisible present moments.
Therefore,
7 The flying arrow does not move.

II

Nāgārjuna's analysis of motion

The arguments presented through the twenty-five verses in the second chapter "*Gatāgataparīkṣā*" of the text *Mūlamadhyamakakārikā* mainly focus on the following three aspects of motion: the locus of motion, that is where does motion take place; the subject of motion, that is who is the mover; and thirdly the time of motion, that is when does motion take place. The arguments ultimately establish that there is mutual dependence between the mover and the place moved upon. And because of this mutual dependence we cannot properly describe motion.

The locus of motion

Regarding the locus of motion, Nāgārjuna's observation is:

Gataṁ na gamyate tāvad agatam naiva gamyate/
Gatāgatavinirmuktam gamyamānam na gamyate/

It is not proper to say that the path which has been travelled (*gatam*) is being travelled upon at present, nor is it proper to regard the path which has not yet been travelled (*agatam*) as being travelled. Further, it cannot also be said that as different from the path already travelled and the one which has not yet been travelled, the path that is being travelled is being travelled.

This argument of Nāgārjuna is based on the classification of the entire path to be travelled into three categories: "that which has already been travelled" (*gata*), "that which has not yet been travelled" (*agata*), and "that which is being travelled at present" (*gamyamāna*). Suppose a person S has to move from one point A to another point C such that there is at a particular moment of time a point B, where S is.

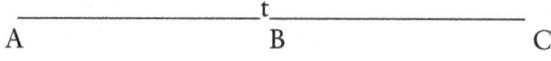

	t	
A	B	C

Obviously for S, the path already travelled is A – B, and the path which is yet to be travelled is B – C. The length of the line is exhausted by (A – B) + (B – C). Nāgārjuna's point is that in none of these positions A – B, B, B – C, can we say the present movement of S to take place. For present movement cannot take place in A – B, the path already travelled, nor can we say that the present movement takes place in B – C, the path which is not yet travelled. To show that the present act of travelling or movement cannot be accounted for in any particular location, Nāgārjuna employs two different types of arguments, which have been referred to by Westerhoff as *property-absence argument* and *property-duplication argument*.

Basic to the property-absence argument is the belief that an individual can be said to possess a property if it is at least conceivable that it lacks that property. For instance when we say that "the table is brown" we can conceive the table and the colour brown as two distinct entities. So we can conceive of a table which is not brown. But the case is different in the context of "the space being travelled upon" and "the act of travelling." In the context of motion, however, Nāgārjuna asks as to how we can attribute motion to the space being presently travelled, since to conceive of the presently travelled space as devoid of motion is not tenable:

> *Gamyamānasya gamanaṁ kathaṁ nāmopapatsyate/*
> *Gamyamānaṁ vigamanaṁ yadā naivopapadyate//*

Moreover, one who accepts that the entity referred to by the expression "*gamyamāna*" (that is, the space being travelled upon) is traveling; one would have to accept the undesired consequence that the expression *gamyamāna* is used with reference to an entity that is not connected with traveling. For to hold this view is to argue that an entity which is devoid of travelling but which is referred to as *gamyamāna* is being travelled:

> *Gamyamānasya gamanaṁ yasya tasya prasajyate/*
> *ṛte gater gamyamānaṁ gamyamānaṁ hi gamyate//*

At the core of this argument lies the idea that, in the context of motion or travelling, the individual depends on the property it instantiates. Since "the traveller" or "the space being travelled" depends on the act of travelling, we cannot analyze the statement "The mover moves" in the same fashion as we can in the statement "The table is brown."

One may justify the common-sense belief that the space that is being travelled at present is the locus of the act of movement or the mover moves, since we can find motion only when there is activity. And such activity is found in the case of the mover or in the case of the space that is being travelled only and neither in the case of the space already travelled nor in the space not yet been travelled. Hence we can say that the act of movement exists only in the case of the space being travelled at present or in the mover. But Nāgārjuna is not satisfied with this position either. He announces that the locus of present motion cannot be said to be characterized by movement. His question to the opponent is how we can speak of the moving / traveling activity of the present space being travelled, for that will lead to the problem of property duplication.

To say that the mover moves, two motions will be implied: the first one being that by virtue of which the individual is considered as a mover and the second one being with respect to which it moves. Actually, in the context of ordinary objects we can distinguish between two different types of properties – the constitutive property and the instantiated property. For example, when we say "brown table," the property of being a table is regarded as the constitutive property and being brown as the instantiated property, and these two properties are distinct. In the context of the presently moving object, the constitutive property is that by which it is moving or that in terms of which it is considered a mover. The instantiated property of the mover is the motion itself. Generally, the instantiated property of an object is different from the constitutive property. For example, when we say "Robinson sings," singing is an instantiated property, for we can conceive of Robinson apart from his singing act. But in the sentence "The thunder roars," we find that we cannot conceive of thunder without its roaring sound, so roaring becomes the constitutive property of thunder. In the context of the sentence "The mover moves," it is seen that the instantiated property and the constitutive property are becoming the same. Hence we are led to the problem of property-duplication in one and the same sentence which cannot be justified.

The opponents may argue that even if there are two different types of movement involved, what is the harm? The reply of Nāgārjuna is that admission of two different acts of motion will lead to the admission of two different agents, as the performing of these two acts of moving, since without the mover the act of moving cannot take place. That is, it will lead to the absurd position that the one act of traveling involved in a particular space's being travelled implies two agents. This absurdity follows because any act for its performance needs an agent. So, when Devadutta is travelling, i.e., performing the act m_1, he is the agent of m_1, but such Devadutta cannot be the agent of m_2 at the same time. Hence to speak of "*gamyamānasya gamanam*" is absurd. It may still be argued that it is possible for an agent to perform more than one act at the same time. For example we often say that while standing in the lawn Devadutta looks at a bird and whistles to himself. Here Devadutta is the agent of three different acts, standing, looking, and whistling, which are taking place at one and the same time. In other words multiplicity of actions does not always imply multiplicity of agents. So why can we not speak of two different acts of movement by the same agent at the same time? The reply is given by Candrakīrti. According to him, the *kāraka*, or the agent in grammar, is not a substance (*dravya*) but a power (*śakti*) that is diversified because of the diversity of the actions. The same Devadutta may perform simultaneously the actions of standing, listening, and whistling, but each of these actions has a different agent: it is not the substance Devadutta which is the same in all these actions but a power – each time different that resides in him.

Who is the subject of motion?

Nāgārjuna proceeds to prove the impossibility of motion by analyzing the notion of the mover. Generally any act presupposes an agent who is the performer of the act. Hence, in the context of the act of moving we have to admit an agent, a subject. But who can be the subject? The subject must be either a mover or a non-mover. Obviously a non-mover cannot be the subject of the act of moving, for that involves contradiction. The problem, however, is that we cannot regard the mover/goer either as the subject. For to say that "a mover moves," with the intention to attribute moving to the mover, involves the possibility of a mover without a moving. For if the mover moves, there are two movements, that by which the "mover" is realized in its capacity as a mover and that which (in fact contingently) moves. That is, "the mover moves" absurdly entails two movers and two movements. The subject, the mover must *a priori* move if she is to be designated as the subject; to this must be added a second contingent movement which is accounted for by the verb and this being a second movement must have a second mover. Hence occurs the paradox.

When can we consider the motion to start?

Nāgārjuna's next line of attack is with regard to the location in the time series when we can speak of the starting of the motion. When motion is completed, movement cannot be initiated, and when there is no motion, movement cannot be initiated either. Where motion is undergone already, movement cannot be initiated. Where then is movement initiated?

The present act of moving cannot take place before motion has occurred, nor is a movement completed where the motion is yet to begin. Hence, how can there be a motion in the case where the movement has not yet started?

Let us explain the essence of this argument in a simple manner. Let us assume that an individual Devadutta during the interval $t_0 - t_1$ is standing at a given location and at some time during the interval t_1-t_2 leaves that location. Let us assume also that there is some time t_x in between $t_1 - t_2$, subsequent to which Devadutta is moving. The question arises: when does Devadutta's moving start? The interval $t_0 - t_x$ exhaustively describes the duration of Devadutta's going for the period. Then since $(t_0 - t_x) + (t_x - t_2)$ covers the entire duration of the analysis, we have to conclude that at no time does Devadutta actually start to go, that is at no time does the commencement of going takes place.

The consequence that follows from this argument is that the gone, the not-yet gone, and the present being-gone-to, as temporal moments, cannot be regarded as anything real. It is impossible to designate these three moments prior to the occurrence of going.

> It is impossible to speak of going actually taking place, however, without this division of the temporal stream, into the three moments of gone-to etc. In other words, a necessary condition of our recognizing the commencement of going is our being in a position to speak of a time where going has ceased, a time where going is presently taking place etc. If we are to succeed in designating the commencement of going, it must take place in one of these three moments – and, of course the not-yet-gone-to may be excluded from our considerations as a possible locus of the commencement of going, since by definition no going may take place in it. Thus the commencement of going must take place either in the gone to or in the present being gone to. This is impossible, however, since neither of these moments may be designated prior to the commencement of going.
>
> (Siderits and O'Brien)

So, Nāgārjuna through this part of the argument tries to uphold two conclusions – first, given the conceptual resources of the triple division of space where motion takes place into the space which has been moved over (a collection of finer special points which are all gone, *gata*), the space which is being travelled now (the point regarded as *gamyamāna*), and the space which is yet to be travelled (the points remaining ahead to be covered, known as *agata*), it is not possible to identify the point where motion actually begins. This analysis shows that the above-mentioned triple division of space with regard to motion is exhaustive, but in none of the spatial points in the triple division can we regard the beginning of motion to have taken place. And this looks awkward since we all recognize that the beginning of motion exists where motion takes place and not where there is no motion. The second conclusion is that it is not possible to define two of the three divisions, namely the space which has been moved over and the space which is yet to be moved, without reference to the point where motion begins and its dual, namely the point where motion has stopped.

This leads us to the paradox. The triple division of space where motion takes place presupposes the beginning of motion. The beginning of motion, in turn, presupposes the triple division of space where such a beginning can be identified. But the beginning is nowhere to be found within the space triply divided, nor can it be found anywhere outside that space. Hence, we have to say that there is beginning of motion and there is no beginning of motion, which is paradoxical.

This paradoxical consequence makes one ponder over the issue as to whether we can then regard the mover to be at rest when we cannot speak of the movement of the mover. Nāgārjuna's reply obviously is in the negative. His point is that "a mover cannot be at rest, nor can the non-mover be. Other than the mover and the non-mover there cannot be any third party who can be said to be at rest." The reason for this is that it is not possible without inconsistency to regard the mover to be at rest, since a mover without movement is inconceivable. Just as a mover cannot be at rest, we cannot speak of anyone who has already moved, who is now moving, and who has not yet moved to be at rest, since movement, cessation, and commencement of movement can all be accounted for with respect to motion.

Inter-dependence of mover and movement

The third variety of argument of Nāgārjuna relies on the interdependence between the mover and the act of moving, between movement and the mover. The concept of a moving object implies the concept of movement by virtue of which the object is regarded as the mover (gantṛ). Again, the concept of movement (gamanam) implies that it is the movement of a moving agent. This mutual dependence of the concept of movement and the concept of the moving agent on each other shows that the two viz. the mover and the movement cannot be considered either as identical or as different. To regard them as identical is to imply that the agent and action are to be regarded as one and the same (ekībhāva). But this is not the case, and we cannot make any particular statement like "this is the act," "this is the doer," etc. We do not identify the woodcutter with his act of cutting. Similarly, we cannot identify the movement with the mover. Again, on the other hand, we cannot regard the mover and movement as distinct (in the sense of independently existent entity). For if mover and movement can exist independently like a piece of cloth and a jar then we can speak of the existence of a static mover, that is, (a mover independent of motion) or of a movement which is not the movement of any mover. But none of these alternatives are tenable.

The main point of the argument regarding the interdependence of mover and motion is that while the concepts of mover and the moved can be regarded as non-identical, neither of them can be said to be self-sufficient or existing by its own; rather, each is dependent on the other.

Ekībhāvena vā siddhir nānā-bhāvena vā yayoh/
Na vidyate tayoḥ siddhiḥ na khalu vidyate//
(*Mūlamadhyamakakārikā*, chapter 2, verse 21)

As the basic principle of Nāgārjuna's philosophy states, those which have dependent existence cannot be regarded as real. Their lack of reality, however, is not the same as that of a hare's horn or the present king of France, since these do not exist either as independent entities, nor do they exist in the empirical level, whereas mover and motion are not real as independently existing entities. What is called mover and what is called motion are all conceptual constructions, and so they have their existence in the empirical level (*samvṛti sat*).

The upshot of the whole analysis is that there is no motion, no mover, and no place to be moved over.

Soteriological interpretation

Nāgārjuna's exposition of motion as contained in the second chapter of the *Mūlamadhyamakaśāstra* can be understood in a double sense in his philosophy. In the first place it is part of the discussion of other entities like agent and action, time, suffering, *samskāra*, *nirvāna*, *Tathāgata*, etc. to show that none of them exist essentially, that is by their own *svabhāva*. And among all these concepts the concept of motion has a special significance, since motion is an example of change and hence is intimately connected with the notion of impermanence (*anityatva*), which is the central theme of Buddhist philosophy. In the second sense, the exposition of motion may be understood as indicating that cyclic existence, or *samsāra*, is nothing but the moving from one birth to another. When Nāgārjuna upholds that the mover, the movement, and the space moved are empty, *nihsvabhāva*, he uses these terms both in their ordinary sense as well as in their soteriological sense, where the mover is that which is born; the movement is the transition from one life to another. Seen from this soteriological standpoint, Nāgārjuna's spirit as contained in the second chapter of his *Mūlamadhyamakaśāstra* is to show that the concepts like birth and death, past, present, and future lives do not have any objective existence in reality but are merely conventionally real boundaries drawn by the human mind. This becomes more prominent to us when we relate it to Nāgārjuna's observation made in the eleventh chapter of the same book: "Where the earlier, the later, and the simultaneous do not exist, then how is there a proliferation of the concepts like birth, ageing and death?"

> *Yatra na prabhavantyete pūrvaparasahakramaḥ/*
> *Prapañcyati taṁ jātim tajjarāmaraṇaṁ ca kim //*
> (*Mūlamadhyamakakārikā*, chapter 11, verse 6)

In short, to point out the distinction between these two types of truth, the conventional truth (*samvṛti satya*) and Ultimate Truth (*pāramārthika satya*), has been the basic objective throughout the text *Mūlamadhyamakakārikā* of Nāgārjuna. Nāgārjuna's analysis of motion is an attempt to exhibit how, by means of logic, we are led to something that would appear to clearly transcend logic. Generally that which transcends logic is thought to belong to the area of spiritualism, an area where science cannot reach. Accordingly a dividing line is drawn between the two poles – the empirical and the transcendental, or the "sacred" and the "profane."[1] When Nāgārjuna in his philosophy repeatedly asserts that we cannot speak of the mover, movement, and the space moved over, he means that none of them can be spoken of in the ultimate sense as something substantive possessing their own beings or as substantive entities, and by that the character of the profane as devoid of one's own being is emphasized.

In other words, mover, motion, etc., all belong to the phenomenal world, to the world of science. For Nāgārjuna, the phenomenal world (*bhava*) alone represents the totality of that which is expressible through language. For example, the various factors alluded to in propositions like "A mover moves," "*samsāra* is no different from *Nirvāna*," the concepts of agent and action, and *nirvāna*, as well as the relationships obtaining between them, constitute part of the phenomenal world. It is only in this phenomenal level for the sake of practical purposes that we need to attribute designations like "movement," "mover," "moved," etc., and all of them are done as relative to each other, and these designations are all superimpositions.[2] At this empirical level, a thing is always without its own-nature (*svabhāva*) and hence is devoid of its own essence. The assumption that a thing has its own-nature turns contrary to our empirical evidences as well as our reasoning. Experience reveals the happening of events only, no own-nature, and no essence. So to apprehend this own-nature we have to give up all sorts of descriptions and attain a stage that is beyond all descriptions and all categorizations.

In the eighteenth chapter of the text *Mūlamadhyamakakārikā*, Nāgārjuna proceeds to identify that which is expressed through language as something non-existent in reality. To express in language what has been grasped by the mind, we have to superimpose something else, namely the impressions of our past experience.[3] Since that which is superimposed is not now existent in reality, what is expressed through language is not existent in reality.[4] This indicates that, according to Nāgārjuna, there are no actual convention-independent entities that correspond to the ostensible referring expressions. Through his dialectic method, Nāgārjuna asks us to purify the given object of thought of all beliefs and dogmas. When such purification is effected, no assertion either affirmative or negative will be possible, and everything will be regarded as *śūnya*. Through the analysis of the motion, Nāgārjuna draws our attention to the fact that truth regarding motion negates linguistic proliferation. When such extinction of linguistic proliferation occurs there occurs the vision of the ultimate truth of emptiness – *prapañcāt prapañcas tu śūnyatāyāṁ nirudhyate*//XVIII.5d. This world of the Ultimate Truth is spiritualism for Nāgārjuna. It is a stage which is ineffable, which cannot be expressed through language. The Ultimate Reality is not constructed with the help of any conceptual constructions, with any sort of linguistic proliferation (*prapañcair aprapañcitaṁ vāgbhir avyāhrtam ityarthaḥ/ Prasannapadā*on XVIII.9) and is free from all sorts of thought process (*vikalpa*), and all types of distinctions –

> *Prapañcair aprapañcitam/*
> *Nirvikalpam anānārtham etat tattvasya lakṣaṇam//*
> (*Mūlamadhyamakakārikā*, Chapter 18, verse 9)

Seen thus, the text *Mūlamadhyamakakārikā* aims to provide a glimpse of the Ultimate Reality in its aspect of *śūnyatā* by getting rid of all linguistic proliferations which are the characterizing feature of the empirical world and which are considered as the root of *karma* and mental defilement. The second chapter of the text which we have discussed here shows with the spirit of negation that from the ultimate point of view we cannot speak of motion. Thus Nāgārjuna's analysis of motion may be considered as a journey from the empirical to the spiritual.

Notes

1 Expressions borrowed from Mircea Eliade.
2 *vyavahāra-satyānurodhena laukika-tathyādyabhyupagamavat tasyāpi samāropato lakṣṣaṇam ucyatām iti. Prasannapadā*on verse 8, chapter 18.
3 *Yadi cittasya kaścid gocaraḥ syāt tatra kincin nimittam adhyāropya syād vācām pravṛttiḥ/ Prasannapadā*
4 *Nivṛttam abhidhātavyam nivṛtta cittagocare/Mūlamadhyakārikā*, verse 7, chapter 18.

Works cited

Primary source

Nagarjuna. *Madhyamakaśāstra of Nāgārjuna with the Commentary Prasannapadā by Candrakīrti*. Ed. P.L.Vaidya. Darbhanga: The Mithila Institute of Post-graduate Studies and Research in Sanskrit Learning, 1960.

Secondary sources

Bhattacharyya, K. "Nāgārjuna's Arguments Against Motion: Their Grammatical Basis." *A Corpus of Indian Studies: Essays in Honour of Professor Gaurinath Sastri*. Kolkata: Sanskrit Pustak Bhandar, 1980.
Burton, D. *Emptiness Appraised: A Critical Study of Nāgārjuna's Philosophy*, Honolulu: University of Hawaii Press, 2001.
Garfield, J.L. *The Fundamental Wisdom of the Middle Way: Nāgārjuna's Mūlamadhyamakakārikā*. New York: Oxford University Press, 1995.
———. *Empty Words*. New York: Oxford University Press, 2002.
Katsura, S. "Nāgārjuna and the Tetralemma (*Catuṣkoṭi*)." *Buddhist Studies: The Legacy of Gadjin M.Nagao*. Ed. J. Silk. New Delhi: Motilal Banarsidass Publishers Pvt. Ltd., 2008.

Mabbett, I. W. "Nāgārjuna and Zeno on Motion." *Philosophy: East and West.* Vol.34.4. (1984): 401–420.

Matilal, B.K. "Mysticism and Reality: Ineffability." *Mind, Language and World.* Ed. J. Ganeri. New Delhi: Oxford University Press, 2002.

Murti, T.R.V. "Nāgārjuna's Refutation of Motion and Rest." *Studies in Indian Thought.* Ed. H. Coward. New Delhi: Motilal Banarsidass Publishers Pvt Ltd., 1983.

Ogawa, H. "Gamyate, Gamyamāna, Gata, Agata: The Mūlamadhyamakakārikā II.kk 1–6 Re-Examined." *The Annals of the Research Project Center for the Comparative Study of Logic.* Vol.2 (2004): 63–75.

Reugg, D.S. "The Uses of the Four Positions of the *Catuṣkoṭi* and the Problem of the Description of Reality in Mahāyāna Buddhism." *Journal of Indian Philosophy.* Vol.5 (1977): 1–71.

Siderits, M. *Buddhism as Philosophy.* Surrey, UK: Ashgate Publishing Company, 2007.

Siderits, M., and J.D. O'Brien. "Zeno and Nāgārjuna on Motion." *Philosophy East and West.* Vol.26.3 (1976): 281–299.

Siderits, M., and S. Katsura. "Mūlamadhyamakakārikā 1 – X." *Journal of Indian and Tibetan Studies.* Vol.9.10 (2006): 129–185.

———. "Mūlamadhyamakakārikā X1 – XXI." *Journal of the Association for the Study of Indian Philosophy.* (2008): 171–221.

Tachikawa, M. *An Introduction to the Philosophy of Nāgārjuna.* New Delhi: Motilal Banarsidass Publishers Pvt Ltd., 1997.

Walser, J. *Nāgārjuna in Context.* New Delhi: Motilal Banarsidass Publishers Pvt Ltd., 2008.

Westerhoff, J. "Nāgārjuna's Arguments on Motion Revisited." *Journal of Indian Philosophy.* Vol.36 (2008): 455–479.

9

INDIAN SCIENCE
AND SEMIOTICS

Some reflections

Sreekala M. Nair

It would be considered an attempt to stretch ideas too far if one dares to draw comparison between modern science and what existed a millennia ago in India – call it Indian science, or Indian logic, or in a more general sense, Indian epistemology. Nevertheless, one would be curious to know the Indian ways of employing inference to the best explanation and compare it with the contemporary theories on scientific explanation. The present chapter is an attempt to explore the scientific insights of ancient Indian philosophy and project it as a parallel way of doing science in contrast with the Cartesian model marked as the epitome of doing science. In order to draw disparities between *doing science* in these two traditions, I shall explore the interrelation between science and semiotics depicted in these two traditions.

Drawing parallels between modern science and its ancient Indian counterpart: problems and prospects

It would raise a few eyebrows if one tries to draw parallels between modern science and its counterpart in ancient Indian philosophy, and of course for valid reasons too. How can there be a shared universe of dialogue between modern science and ancient and medieval Indian philosophy, the latter being born of a distinct civilization in a different historical setting? While addressing this problem, first of all we need to realize that philosophy of science is first and foremost philosophy. And those who have even a peripheral understanding of contemporary philosophy of science would accept that it draws insights from ancient and medieval Western philosophies, which in some way suggests that some of the basic philosophical questions that are fundamental to the understanding of the nature of science continue to make sense even today (Sarukkai 2005, 158). Having highlighted the significance of ancient Greek philosophy to the contemporary philosophy of science, I would like to take a step ahead and bring home the fact that there is more commonality between ancient Greek thought and ancient or classical

Indian thought than between Greek thought and its modern descendants. The questions both these ancient thinkers raised were the same, and despite the fact that the methods they engaged in were different, I would argue that there are a lot of unidentified similarities between them. In fact some of the basic questions that philosophy of science addresses to this day, such as, "How do we move from observations to generalizations?" "How can we distinguish between accidental generalizations and necessary relations?" and "What exactly is the role of language in describing the world?" etc., were also the common concerns of both ancient Greeks as well as ancient Indians (ibid.). Arguing in this line it will be clear that science and its unique method that distinguishes it from other knowing enterprises were not alien to Indian thinkers. The present chapter would attempt to establish that Indian philosophy has been employing scientific method since ancient times and, what's more, even their logic and epistemology are structured in consonance with scientific rationality.

A number of contemporary Indian philosophers, like Matilal, Sibjiban Bhattacharya, Jonarden Ganeri, Sundar Sarukkai, and others, have argued that the image of science in India is pretty different from that of the Western tradition. While the West insists that science be logical, Indian thought structure places logic in science. For the Western world, mathematics/logic functions as the core or basic science, the foundation as it were, upon which the edifice of the knowledge system is being erected. Quite different from this, Indian knowledge structure puts forth an alternative model where language is being placed as the foundation upon which empirical science emerges as a natural product, as suggested by Sarukkai. And it is within this empirical science we find the unfolding of logic as a separate discipline.

Semiotics in Indian logic

The notion of sign plays a significant role in the unfolding of human thought. A sign by definition is something which has the potency to represent or stand for something else, a mark indicating the presence of something else. In empirical life, we have examples like that of a picture representing an object or a person. One is here reminded of a popular anecdote from the life of Swami Vivekananda, wherein he defends the notion of idol worship using this theory of representation. The point that emerges here is that there is a relation that runs between the sign and the signified, either causal or nomological. A closer look at Indian philosophy would reveal that this science of signs, called semiotics, runs as an undercurrent in almost all theoretical exercises of Indian knowledge system. That the concept of sign plays a central role in Indian logic/epistemology/science as well as in modern science is known to all. Therefore, by exploring the uniqueness of the relation between the sign and the signified in Indian knowledge analysis, one would be able to strike at the fundamental disparities between Indian and Western sciences.

In Indian epistemology, analysis of sign happens to be a major chapter in inference. Though undoubtedly the analysis of the nature of sign in India predates that of the West, a systematic study of the relation between sign and signified and its role in inference emerged only with the Buddhist logicians. In fact, as Sarukkai observes, Dignaga turned the question of logic into a question of semiotics (Sarukkai 2005, 53). Dignaga, quite unlike Nyāya, focuses attention on the third member in Nyāya 's five member syllogism, *wherever there is smoke there is fire*, and seeks justification for that. He seems to be obviously in search of the proof for certainty with regard to inductive inference (Sarukkai 2005, 53–55). In fact, Dignaga offers a structure of inference based on the nature of the sign, which we could call as a proper analysis of sign, wherein he offers the triple nature of sign as follows:

It should be present in the case under consideration (*Paksa sattva*).

It should be present in a similar case or a homologue (*Sapaksa sattva*).

It should not be present in any dissimilar case, or a heterologue (*Vipaksa asattva*).

(Matilal 2002)

In the traditionally sought smoke-fire example, smoke is the sign for fire. In order to qualify smoke as a valid sign leading to prove the presence of fire, we need to examine whether it fulfills the three requirements stipulated above. In the case under consideration these three conditions, viz. *paksa sattva*, *sapaksa sattva*, and *vipaksa asattva*, are indeed fulfilled by the sign here, namely, smoke. By stipulating *sapksa sattva* and *vipaksa asattva*, Indian logicians are trying to establish a necessary relation between *sadhya* and *hetu*. They did not however phrase it in terms of necessity using model language and discussion on the types of necessity, viz. the empirical and logical, has never emerged in Indian logical debates. The occurrence of the sign and signified together is seen as illustrating a relation between them, a relation of invariable concomitance, or *vyapti*. The need for establishing *sapaksa sattva* emerges in view of the inductive leap one will have to take with regard to empirical entities. Since empirical generalizations cannot be verified in principle, all that we can do is to increase its probability rating by finding more similar cases. Indian logicians felt that sheer verification alone would not suffice. Hence *vipaksas* or heterologue instances are made use of, the so-called negative examples.

The importance of the third condition is based on the observation that a single counter instance would be able to disprove the thesis at hand. Gangesa has beautifully defined *vyapti* as *vyabhicarirahita sahacarisambandhah*, confirming the co-absence being crucial to the establishment of the invariable concomitance. Here one cannot miss the striking similarities between Karl Popper's theory of falsification and Gangesa's renewed definition of *vyapti* (many contemporary Indian philosophers like Matilal, Ganeri, and Sarukkai have drawn this comparison). In his theory of falsification, Karl Popper has called upon the scientific community to focus on falsification of a hypothesis rather than confirmation of it. He has popularly argued that if a hypothesis is open to being shown wrong, then it is a better scientific hypothesis. Well, this reflection should not be interpreted as one drawing comparisons between Gangesa and Karl Popper, for there are significant differences between Gangesa and Karl Popper with regard to highlighting the negative instances.

Sign as reason and evidence

We would go wrong if we confine signs to just material signs in the Indian instance, as here sign and reason are interchangeably used. The previously stipulated conditions of sign, we may take note, are equally applicable to reason/evidence as well. Examples abound in Indian logic where relations between the non-material sign and the signified get discussed, *knowability* and *nameability*, *producthood* and *non-eternality* being some of them. While Western scientists developed and explored the relation between concepts like force, mass, and acceleration, Indian philosophers as well as scientists explored the relation between *knowability* and *nameability*, *producthood* and *non-eternality*, etc. Though thinkers of these two traditions concentrated on different aspects of reality, both were concerned with reasserting the validity of these relations as well as ways to reach certainty or something closer to it.

Having highlighted the commonalities with regard to scientific method between Indian and Western traditions, let us also note two essential differences which would help us understand how Western scientific reasoning parts ways with Indian logical reasoning. First, in Western science concepts are open to rectification based on observation. Concepts in science in the West are found being used to define and create fresh concepts, and they all together stand in a coherent relation with one another. But this is unfamiliar for the Indian mind. Also the desire for exploring and exhausting the virtual universe, probably due to its heavy metaphysical footings, is absent in Indian theories. Second, in modern science the establishment of the relations between various basic constituents of the universe, like matter, energy, force, etc., is found used in innumerable contexts; for instance, Newton's law of motion finds its applicability in hundreds of situations. In other words, theory is guided towards its practical applications; science is meant for technology. But in India the situation is pretty different; the task of the logician seems to end when the analysis of the relation between eternality and producthood of sound stands analyzed. A step towards technology making use of this valid relation to build something else is alien to the Indian mind (Sarukkai 2005, 162).

The semiotic character of Indian logic is often overshadowed by the all-pervasive inference here. In order to explore the semiotic character of Indian logic, it is essential to analyze semiotics and its role in logic. The term sign is derived from *signum* meaning proof or clue or symptom. The common semiotic space shared by these meanings ratify the way sign, reason, and evidence are used interchangeably in Indian logic. It is significant for us to note that Indian epistemology considers sign, reason, and evidence as synonyms. Smoke functions as the evidence for believing that there is fire, and the presence of smoke is also the reason for coming to the conclusion that there is fire (Sarukkai 2005, 54). As Sarukkai observes,

> The idea of reason is already inherent in the meaning of sign of something else and it is conceivable that there is a reason for this connection: the question for the Indian logicians consisted in knowing whether the sign really stood for the thing which it supported to stand for.
>
> (Sarukkai 2005, 167)

Signs: arbitrary or natural?

Signs, as we know, are of two types: arbitrary and natural. While the West lavishly used arbitrary symbols, especially in their formal sciences, the same is not practiced here in India. At the same time we will also have to take note of the fact that the use of arbitrary signs was not unknown to them. Dignaga's *apoha* doctrine and Naiyayika's referential theory of meaning point towards this fact. Naiyayikas and Buddhists in their philosophy of language maintain that the word–world relation is arbitrary, while Mimamsakas and Grammarians hold that the world and its linguistic representation is determined. Interestingly, the former group happens to be the one which has significantly contributed to logic. Now the question is, if they can afford to accept the arbitrariness of world–word relation, why not the same in logical signs too? A straightforward answer to this would be hard; nevertheless, it seems that according to Indian logicians, while there can be a sign which can, in principle, stand for anything, there would be a subset of these signs which has a special, natural relation with the signified. This subset consists of signs that are used in inferential exercises. The synonymous use of sign/reason/evidence might have been one motivation for the formation of this subset.

Arbitrary symbols are of two types: one which has a corresponding semantic world, and the other which does not carry a space of meaning. Linguistic symbols belong to the former category; though they are arbitrary symbols, they are filled with meaning. But symbols in formal sciences like logic and mathematics do not have any corresponding semantic world to which it may get attached. While discussing the scope and merit of arbitrary symbols, which do not have semantic extension, we need to remember that

111

the use of arbitrary symbols did cause significant advancement of Western formal sciences. However, Indian intellectual tradition, for a variety of reasons, has turned down the possibility of arbitrary signs in inferential reasoning; for Indian logicians, the nature of a valid sign is determined on a natural relation, largely causal.

If we glance through the development of logic in the Western tradition, from Aristotelian logic down to current propositional and predicate logic we find ample use of arbitrary signs. For instance, in the Aristotelian syllogism, the premise "All men are mortal" can be formalized, letting "S" stand for "men" and "p" for "mortals," and we can represent the premise thus: "All S is p." Note that here the use of signs, like in the case of our use of words for objects, is quite arbitrary; obviously there is no *vyapti* or pervasion between S and for what it stands for, namely, all Greeks. For Indian logicians, this move is not permissible. They seek an answer to the question, "What is it that permits us to use a sign to stand for something else?" The answer to this question would form the very basis upon which the symbology under consideration is made possible. We could also consider this as a challenge posed to modern Western logicians from their Indian brethren to clarify the presupposition inherent in their very act of symbolization. Indian logicians have resisted the use of arbitrary signs, for the moment we start using arbitrary symbols we move away from cognitive/ scientific inferential exercises and shift to an altogether different realm, the realm of language (Sarukkai 2005, 170). An awareness of this basic approach of Indian logicians would also warn us against over-enthusiastic attempts to symbolize Indian logic in certain quarters along modern or Western lines.

The fundamental question behind any scientific theorizing, be it modern Western or Indian, is how to undertake the journey from empirical observations to a hypothesis and then to theory. The first step towards this would be converting an observation, which is cognitive, into linguistically intelligible terminology. This first move, from observation to putting it into language, itself initiates the process of theory-making. By the use of language we have already started moving from the phenomenal world to the world of signs and symbols. Once we make this shift, the rest of the process is a continuation, adding sophistication to the methodology. In brief, the theory-enmeshed world of signs, from lesser to more complicated ones, is a product of human thought. Philosophers of mathematics such as Frege, Boole, and others believed that the idea of sign was a great discovery and that the laws of sign are visible expressions of the laws of thought. Cassirer, yet another mathematician belonging to the same tradition, viewed that the structure of science rests on the *logic of things*. By the *logic of things* he refers to the material concepts and relations, which very much includes within its purview the logic of signs (Sarukkai 2005, 172). This suggests that, in scientific reasoning, as we proceed from observation to theorizing we eventually give up any dealing with the material entities and shift to the realm of semiotics.

In other words, concepts of science are no more imitations of existing things but only symbols connecting the reality in a functional way. In brief, concepts are semiotic.

That the concepts are semiotic is something that demands explanation: concepts in some sense are the objective realities already present in the phenomenal world and are therefore discovered during our empirical observations. The concepts describe the observation and abstract the structure of them and thereby play a mediating role between observation and theory. If concepts are to play this mediating role, they cannot be part of either the phenomenal world or the world of pure abstractions, the world of theories, but they are semiotic, belonging to a world of semi-abstraction, enabling the human mind to transcend the empirical and proceed to abstract theorization. Little surprise that thinkers like Cassirer felt that the concept of symbol is the actual focus of the intellectual world, the only trouble being that Cassirer was over 1,000 years off the mark. Indian logicians, especially Buddhist logicians like Dignaga, have done detailed analysis on the nature of signs centuries before him (Sarukkai 2005, 173).

Semiotics as part of scientific realism

It is to the credit of science that it opens up the possibilities of understanding unobservable entities by our normal perception. With the help of sophisticated instruments we are able to widen the vision of our perceptual world. Thus, there are two modes of perception in science: the direct and indirect. Now in the world of instrumental perception we are, in fact, interpreting what we perceive because all that we have are signs referring to something else. Scientific realism is based on the axiomatic presupposition that our world is filled with entities unobservable, and we access them only through signs. As has been said,

> Science cannot perceive the hidden forces of nature, the order in nature that controls the world. It has to infer all the elements of this order. This extends from inferring objects like electron to the fundamental law of the world. Given its immersion in its semiotic mode, knowing well that only signs are accessible to us, it is not surprising that science is fundamentally concerned with knowing what these signs really stand for.
>
> (Sarukkai 2005, 181)

Now here is something that requires our attention. All these signs are natural signs; they are signs that stand in natural relation with the signified. This reveals that within the larger world of Western science, signs go natural just like in Indian science/epistemology.

Theory dependence of natural signs in science

The theory-laden nature of signs in science is in fact, a problem that finally reaches from science to the camp of philosophy. The problem can be best explained with the aid of the cloud chamber experiment. The cloud chamber has an electromagnetic field. The experiment output can be generated as a photographic representation which has a lot of lines on it. One of these lines corresponds to the track of an electron. There is no direct perceptual content in the scratchy line, and we really cannot infer how an electron looks based on this line. However, this line indicates the presence of an electron. The crucial point is, if we do not hold a theory that tells us how an electron would behave under the action of electro-magnetic field, we would not be able to interpret a particular line as being indicative of the motion of an electron. Now this line is a *sign*, which specifically stands in need of a theory to get connected with the signified, here the electron. In other words, before we see the line indicating the movement of electron, we should have a theory describing how the sign of an electron will look and under what conditions it will be found. This heavy theory dependence is a characteristic of science, and often experiments are designed to merely verify a given theory. It would be interesting if, before addressing the question how much a theory contributes to what I think I am observing, we first visit our ordinary perception. A close look at our perceptual experience would reveal that, even in ordering perception, we seem to perceive more than is available to the senses. As Naiyayikas would say, while perceiving an individual object we also perceive the class characteristics, *Jati*, which is something additional to what is given or presented before us. In other words, our perceptual objects necessarily pass through certain conceptual categories. Our interest in presenting the theory-ladenness of scientific observation is to make visible the scope and extension of the semiotic world of science, wherein through theoretical reflections science describes a net of signs that will point towards the existence of scientific objects.

Yet another act of reason related to the observation is the ability to see patterns in nature. Scientific generalizations are made possible by noting these patterns. Patterns and regularities have to have a cause since nature shows an inherent tendency for disorder. So any order is an indication, a sign of hidden forces that shape that order. Here again we have a point of similarity between modern science and Indian logical tradition. Both the traditions note regularities in nature and proceed to construct a discourse to explain these regularities. And both are equally concerned about the level of certainty to which they can raise their theories.

Semiotics for scientific explanation

We have already seen that semiotics is an important theme that can help us explore Indian logic along new paths. The emphasis on natural signs against arbitrary ones is also indicative of the presence of an explanatory structure. In what follows we shall see how Indian inferential arguments do a job quite similar to what the model scientific explanation does, and thereby once again reiterate the fundamental similarities between modern science and its Indian counterpart.

Many contemporary Indian philosophers have drawn parallels between the Nyāya five-member syllogism and Carl Hempel's deductive nomological (DN) model of explanation. What follows is the sum and substance of the Nyāya five-member syllogism:

Wherever there is smoke there is fire
Like in the kitchen
There is smoke on this hill
Therefore there is fire on this hill.

That this syllogism fulfills all the requirements of Hempel's DN model would be evident if we cross-match it with the conditionals of the latter:

The scientific explanation must be a valid deductive argument
The explanans must contain at least one general law actually needed
in the deduction
The explanans must be empirically testable
The sentences in the explanans must be true.

The similarities with the DN model should make us realize that the five-step process is primarily a deductive process, just like the DN model. Both DN and the Nyāya model use examples as part of their structure and this is the strength of a scientific explanation. The fact that the five-membered syllogism is called *inference for others* (*paraarthaanumana*) also suggests that this is an explanatory model accounting for how a particular inference is made. To sum up, the larger Nyāya project indicates strong correlation with the fundamental aspects of scientific methodology of the West (Sarukkai 2005, 196).

To cut short a long argument, one can say that in the West logic came first and then explanation. In the Indian traditions, the scientific explanation came first, and logic was part of the larger structure of explication. Also, Indian logic differs from the Western tradition in that here logic is not exclusively confined to the formal laws of thought but works on a necessary combination of the laws of the world and the laws of thought. We may also note that quite unlike in the West, Indian logic makes no commitment to empty terms, and hence logic is always the logic of the empirical. This brings inference and explanation naturally together. Hence, we may understand that in Indian logic epistemology and science are all members of a joint family, quite uncomfortable with the idea of a separate existence. What is more, all these branches of knowledge had a single common goal: *nisreyas*!

Works cited

Matilal, B.K. *Perception: An Essay on Classical Indian Theories of Knowledge*. New Delhi: Oxford University Press, 2002.
Sarukkai, Sundar. *Indian Philosophy and Philosophy of Science*. PHISPC Vol. 15. New Delhi: Centre for Studies in Civilizations, 2005.

10

INTEGRAL NON-DUALISM AND MODERN SCIENCE

Some reflections

Debabrata Sen Sharma

A bird's eye view of the different theories of non-dualism (*Advaita*) postulated by different schools of Indian philosophical thought show four distinct projections in them. Looking from the historical perspective, the first and earliest projection of the concept of non-dualism was made by the Mahayana Buddhists, represented by the Yogacara and the Madhyamika schools. They advocated the theory of non-dualism based on the negation (*Nisedha*) or the denial of the existence of the multiplicity of the world on the ground that it is merely false superimposition (*Adhyaropa*) on the substratum (*Adhisthana)*, the perennial or the non-dual stream of consciousness (*Vijnana Santana*) which is incapable of being comprehended or described in positive or negative terms or both. The negative description of the substratum given by the Madhyamika Buddhists was technically named as *Sunya* (literally, vacuity). Since the momentariness (*Ksana Bhavigavada*) is one of the cardinal doctrines of Buddhist thought, everything, including the underlying substratum (*Adhisthana*), i.e., consciousness, is momentary or in a state of flux. The multiplicity of the world, which is also undergoing mutation, is merely superimposition upon the substratum, consciousness existing in a state of flux. Since our senses, the instruments of cognition, are also undergoing change, we are unable to perceive the flux going on within us or outside us in the world of objects. We imagine them to be constant and existent, which is not true and only a delusion (*Bhranti*). The Buddhists therefore are labeled as subjective idealists. As a matter of fact, their doctrine of non-dualism appears to be a misnomer and only a pseudo-non-dualism.

Gaudapada, the founder of a distinctive school of Vedanta, which later came to be known as the Advaita Vedanta, propounded an apparently similar kind of the doctrine of non-dualism which, however, cannot be called pseudo-non-dualism. This is because the non-dual Reality conceived by him to serve as the substratum for the superimposition (*Adhyaropa*) of the multiplicity of the world is not ever changing. It is in fact existent as the immutable Reality on which the percipient subjects like us make the superimposition of the multiplicity of the world. Gaudapada's concept of non-dualism was adopted and further reinforced by the Adi Shankaracharya, the grand disciple of Gaudapada, and his followers. Adi Shankaracharya postulated the second kind of doctrine of non-dualism, the non-dual substratum (*Adhisthana*), is real and existent, but the multiplicity of the world-manifestation is not, it is only a superimposition on the substratum, Brahman. The multiplicity of the world has only the phenomenal existence, it has no numeral existence at all. By denying the existence of the multiplicity of the world in the Brahman, the non-dual one, Shankaracharya asserted that the non-dual Brahman can be reached only through the negation of existence of everything that is not Brahman but is perceived by us in the world (*neti-neti-vada*).

The schools of vedanta that are inclined towards theism, namely those of Ramanuja, are Vallabhacarya, which propagates their kind of non-dualism, which is called the Visitadvaitavada and Suddhadvaitavada, respectively. The kinds of non-dualism they postulate are not "pure," as these appear to be "adulterated" by theistic ideas and, as such, these appear to be influenced by the Bhagavat tradition.

Ramanuja, as we have already mentioned, propagated the doctrine of non-dualism, technically called the qualified non-dualism (*Visistadvaita-vada*). According to him, the ultimate Reality is the function of *cit, acit*, and *Isvara*. By *cit* he means limited beings or *jiva*, while *acit* signifies the Prakriti or Maya, the principle of insentient material power (*Jadasakti*); *Isvara* was the unifying principle of the two *cit* and *acit* fused together as it were. Ramanuja's conception of the *Isvara* is akin to *Sagura* Brahman (qualified Brahman). He appears to have taken the cue from the description of the Brahman seen in *Svetasvatara Upanisad* (I, 8), which reads thus: *Samyuktametat Ksaram aksaram ca vadhyata bhoktribhavat, jnatva devam mecyate sarvpasairh.*

Vallabhacarya conceives the Brahman in almost the same way. According to him, the Brahman is the repository of all attributes, yet He himself is free from them. He is *Nirguna* (devoid of all attributes). He created the world with the help of His *Sakti, Mayasakti*, out of his free will as a part of his divine play (*lila*). Both He and His creation are real, but this does not affect His non-dual character in the heart. In this context, it may be mentioned here that Vallabhacarya's concept of non-dualism, his ideas about the nature of the Brahman and the world, bear striking resemblance to, as we shall see, those projected by the Kashmiri Saivites. Maybe Vallabhacarya, who lived in the fifteenth century, was influenced by the spiritual thought-projections of Kashmiri Saivites expressed by them in the period between the eighth and the twelfth centuries AD. Vallabhacarya talks about the *Para* Brahman (supreme Brahman), *Aksara* Brahman (the immutable Brahman), and the Brahman manifesting itself as the world, called the *Ksara* Brahman by him. He tries to integrate them, the *Ksara* and *Aksara* Brahman, into the One Brahman, the Supreme Brahman, as *Para* Brahman in his philosophy.

Against the background of these different kinds of non-dualism described above, let us now focus our attention on the fourth kind of non-dualism, ethnically called the integral non-dualism (*Akhand Advaita Vada*) projected by the Advaita Saivism of Kashmir, popularly known as Kashmiri Saivism. This kind of non-dualism envisages the existence of everything in the non-dual One, technically called the *Samvid* or *Caitanya*. The Kashmiri Saivites believe in the integration of everything with the eternal existent Reality. They seem to make the simultaneous affirmation of the multiplicity and the One possible by making it all-embracing and all-inclusive.

In other words, they conceive the ultimate Reality to be the integral One (*Akhanda*) in which the infinite multiplicity and all their mutual differences "dissolve" or lose their destructive identity. All contradictions arise and exit on the level of our finite intellect. Once the level of finite existence is transcended by us, and we are able to enter into the realm of infinite consciousness, we find that all contradictions disappear automatically. One and many, the absolute (*Anuttara*) and the universal (*Visvatmaka*), constitute the two facets of the ultimate Reality, *Samird,* indicative of its fullness-nature (*Paripurna Svabhava*).

I would now like to draw attention to some other instances of integral vision (*Akhanda Dristi*) reflected in the metaphysical thought-projections of Kashmiri Saivism. One instance of this vision is seen in the manner the Advaita Saivites integrate the two apparently divergent components of the world manifestation – the sentient (*Cetana*) and the insentient (*Acetana*) – by describing the ultimate Reality, *Samvid* as *Prakasa-vimarsamaya*. Let me elaborate this problem of integration confronting all non-dualist philosophers and the solution given by the Kashmiri Saivites.

We are all aware that the world is comprised of two distinct categories of existent things, the sentient (*Cetana*) represented by the self-conscious subjects and the insentient material objects (*Acetana Jada Visaya)* that are inert and dead. All protagonists of the doctrine of non-dualism find it difficult to explain the divergence between the two components in creation and try to integrate them in order to uphold their belief in the philosophy of non-dualism (*Advaitavada).* The Savikara Vedantins try to get over this hurdle by regarding the world manifestation to be phenomenally real but unreal looking from the perspective of the Brahman, the ultimate Reality. The experience of subject and object (*Asmad-yusmad Pratyaya)* is an offshoot of *Adhyaropa*, i.e., the superimposition on the Brahman of what appears the multiplicity of the world. But the Kasmiri Saivites, who regard the world as real (*satya*) and existent, overcome this logical hurdle by defining consciousness (*caitanya* or *samvid)* as *Prakasa-vimarsamaya*. The term *Prakasa* literally means the self-effulgent light that always shines and reveals itself (*Svayam Prakasa Jyoti*). It symbolizes the immutable being – nature in the *Caitanya* or *Samvid* – somewhat similar to *Sat* and *Cit* nature of the Brahman described in texts of Savikara Vedantins. The term *Vimarsa* denotes the

119

ever-pulsating Sakti (*Spandana Sila Sakti*) innate in the *Caitanya* (consciousness) that renders it always remaining self-conscious or self-aware. It is for this reason that *Caitanya* is very often described as the *Samvid* (experiencing principle) in the Saiva texts. By defining *Caitanya* this way, the Kashmiri Saivites find it easy to integrate the sentient creation with the insentient one. Applying the above mentioned definition of *Caitanya* or *Samvid* to the two components, sentient and the insentient, in the world we notice manifestation (*Sphutarupa*) in the insentient material things as responsible for their being perceived by us in existent form (*Sadrupa*), while the *vimarsa* aspect of *Caitanya* lies latent in them. This is precisely the reason that the insentient or material objects (*Jadavatn*) are perceived by us as not self-conscious or self-aware. Therefore material needs to be revealed to us in cognition by the light of the consciousness (*cit-prakasa*); only then we are able to cognize them.

The second instance of Kashmiri Saivites integral view in their thought: projections in the field of metaphysics is their conception of the Divine Sakti of Supreme Reality (*Parasamvid* or *Parama Siva*) functioning in three different forms while manifesting the cosmos or world. The different forms are *Mahamaya Maya* and *Prakriti*, which function on the three stadia in the world creation.

As mentioned the *Samvid* according to Kashmiri Saivism is endowed with Sakti- *Vimarsa Sakti* constituting its very nature. It is held by the Kashmiri Saiva school that when the ever-vibrating divine Sakti remains completely fused with *Samvid*, She reveals its transcendent (*Visottirna*) immutable nature. But when the same Sakti pulsates somewhat separately from the *Samvid*, she reveals its immanent from (*Visvatmaka*) as the world manifestation. Some Kashmiri Saiva texts go to the extent of describing the manifestation of the world as an "expansion" of the *Samvid* in the form of its Sakti (*Ava Sakti Sphara*).

It is said that the manifestation of the world takes place in three different forms on the three stadia (*Bhumi*) in creation. On the top most stadium, the Divine Sakti functions in the pure form as consciousness power (*Cit Sakti*), technically called the *Mahamaya*. The sphere in which the Divine Sakti functions as *Mahamaya* is named as the *Suddha Adhva*, the realm of consciousness-power operating in pure form. Two characteristics viz. non-dualism (*Advaita*) and universality (all-pervading form) dominate all levels of creation within the realm of the *Mahamaya*. The same divine Sakti functions in the form of *Maya Sakti* on the second stadium lying just below the sphere of *Mahamaya*. Since the *Maya Sakti* is described in the Kashmiri Saiva texts as the power of obscurantism (*Tirodhana Sakti*), she is held responsible for "obscuring" or hiding the true nature of the *Samvid* like clouds hiding the sun in the sky. She is not alone in obscuring the "face" of the *Samvid* but is assisted by the progeny of five sheaths, technically named as *Kancukas*, in enwrapping the *Samvid*. These sheaths are said to lie in the sphere of *Maya*. It is in this sphere that the seeds (*Bija*) of duality are sown which, later produce multiplicity and difference. The same Sakti taking the form of *Prakriti Sakti* on the third stadium in creation is held responsible for the manifestation of the world in gross form and the multiplicity that we experience in every sphere in the world.

From this description of the nature of world phenomena occurring in Kashmiri Saiva texts, it is obscurism that veils everything which we perceive in the world whether in the form of sentient self-conscious subjects or insentient material objects. But all are manifestations of the consciousness force (*Cit Sakti*) of the *Samvid* which constitute the core of their nature. We, normally staying on the mundane plane, are not able to experience the "Play of Sakti" that the world manifestation symbolizes, but the Advaita Saiva philosopher considers this to be true and real as testified by the *Sadhakas* realizing their Siva nature. They do not experience the world-manifestation, material forms consisting of multiplicity of conscious subjects and material objects, but find it transformed as the wave (*urmi*) of the divine Sakti arising in the ocean of their conscious nature. They articulate their supreme spiritual experience by uttering "*Sivo'ham Sarvo'yam mamaiva vibhah*" (I am Siva, all this world is the expression of my divine glory or nature).

This spiritual experience by *Sadhakas* is corroborated by the findings of physicists in modern times that matter and energy are not two different entities and that which is perceived by us as the solid mass of matter is only energy in different form. They have also discovered a reservoir of energy even in an atom of matter, which they are trying to harness for improving the quality of human life.

As has already been shown in the foregoing paragraphs, the Advaita Saivites are well known for looking at everything in the world from an integral perspective. They have tried to integrate the sentient and the insentient components in creation by adopting integral vision. Looking at the phenomena from an integral perspective, the physicists led by Einstein have also tried to explain the laws governing the physical world by formulating unified field theory. It appears that the similarity in the perspective of the two different disciplines is trying to understand and explain the nature of world- phenomenon, which is anything but superficial.

11

PRINCIPLES OF PLANT TAXONOMY

A fresh insight into the ancient Indian methodology and philosophy of naming and classifying medicinal plants

Sampadananda Mishra

Proper nomenclature and classification play an important role in the systematization of any branch of knowledge. In this regard the ancient Indian *rishis* and *acharyas* have shown much transparency in their scientific observations. To them, to name was to touch the essence of the thing or object named. They could really enter into the soul or the consciousness of the thing or the object and then give the name as per their experience. We find a clear reflection of this in the names of the plants as they appear in various texts of Ayurveda. From the various names given to one plant one can truly understand not only the various morphological characteristics of that plant but also the special medicinal properties that the plant has. This is still a mystery as to how the ancient Indian *Vaidyas* or medical scientists could discover the exact property of a plant and its multidimensional aspects when there was no facility compared to what empirical science has today. This chapter brings a fresh insight into this aspect and throws light on the ancient Indian methodology and philosophy of naming and classifying the medicinal plants.

Sir William Jones, the founder-president of the Asiatic Society of Bengal, once wrote: "I am very solicitous to give Indian plants their true Indian appellations, because I am fully persuaded that Linnaeus himself would have adopted them had he known the learned and ancient language of this country" (*Botanical Observation of Select Indian Plants*, published in *Asiatic Researches, Vol.4 237–312, 1795*).

Unfortunately, this suggestion has not received serious attention even from the Indian botanists. The ancient Indian naming of plants, as found in the various texts dealing with the subject, is an extremely interesting topic which gives us an indication not only of the geography of India but also highlights that our ancestors had a keen sense of observation. In the texts dealing with the names of the plants, we find that a plant is referred to by many names. A proper analysis of all names of a particular plant reveals that each name describes a particular feature (*svarupabodhakanama*) or a specific quality/property (*gunabodhaka-nama*) of that plant. Plant names are coined on the basis of varied criteria. Before I deal with the details of these criteria, let me give you an idea about the various classifications of the plants done by various sages and *acharyas* of ancient India.

Classification of plants

The ancient Indian classification of plants was based on three major considerations, namely, *udbhida* or botanical, *virechanadi* or medicinal, and *annapanadi* or dietic. I am not sure how practical this classification is from the present standpoint, but this classification provided a working basis for botanists, physicians, and agriculturists of India for a very long time.

It is obvious from various Sanskrit texts that the ancient Indians did recognize different kinds of plants based on their habits. They also recognized plants that bear flowers and fruits and those that do not. Thus, Manu, the famous author of *Manusmriti*, classified the plants into eight different types (*Manusmriti*, 1, 46, 47, 48). They are:

1 *Oshadhis*, or those that bear abundant flowers and fruits, and wither after maturing, e.g. rice and wheat.
2 *Vanaspatis*, or those that bears fruits but no evident flowers, e.g. Auster fig (*udumbara*).
3 *Vrikshas*, or those which produce flowers and fruits, e.g. Neem (*nimba*).
4 *Guchchhas*, or bushy herbs of various types, e.g. Jasmine (*mallika*).
5 *Gulmas*, or succulent shrubs of various types, e.g. Nerium.
6 *Trinas*, or grasses of different kinds, e.g. Cuscus grass (*usheera*).
7 *Pratanas*, or creeper with stems spreading on the ground (procumbent and decumbent), e.g. *prasarini*
8 *Vallis* or those which twine round or climb a tree or a support. e.g. *guduchi*.

Acharya Charaka gives a different classification of the plants (*Charakasamhita, sutrasthana*, 1. 36, 37). According to him, trees that bear fruits without flowers are *vanaspatis*; trees that first bear flowers and then fruits are *vanaspatyas*; herbs with spreading stems are called *virudhas*, and they are further divided into two classes – creepers, or *latas*, and herbs with succulent stems and shrubs, or *gulmas*. Those herbs that wither after maturing are *oshadhis* and are further divided into two groups: annuals or perennials bearing fruit, and plants that wither away after maturing and without fruiting, e.g. grasses like the Bermuda grass (*durva*).

Acharya Sushruta lays down an almost similar classification of plants (*Sushrutasamhita, sutrasthana*, 1. 23). According to him, there are four varieties of plants: trees which bear fruit without blossoming (*vanaspatis*); those that bear both fruits and flowers (*vrikshas*); shrubs and creepers that trail (*virudhas*); and plants which die with the ripening of their fruits (*oshadhis*).

Apart from the above-mentioned *acharyas*, Udayanachary, the author of *Kiranavali* (Bibliotica Indica 256); Prasastapada, the *Vaisheshika* commentator (Vizianagram Sanskrit Series, Vol. IV 28); and Amara Simha, the author of *Amarakosha* (*vanaushadhivarga*) have dealt with the botanical classification of the plants in great detail. The *Bhagavata Purana* also deals with the classification of plants in its third *Skanda*, tenth chapter, verse No. 19.

In the ancient Indian medicinal treatises the plants were studied mostly in relation to their medicinal properties or values. Thus we see that Acharya Charaka (*Charakasamhita, Sutrasthana*, IV) classifies the plants primarily into two divisions, 1) purgatives or *virechanadi*, and 2) astringents or *kashayadi*. According to him there are six hundred purgatives and five hundred astringents. Acharya Charaka puts these herbs under various groups or *vargas* like *jivaniya*, or those prolonging life; *lekhaniya*, or those reducing corpulency or thinning the tissues; *dipaniya*, or those promoting appetite and digestion, etc. This classification of Acharya Charaka is based on the finer properties of the plants (Majumdar, *Plants and Plant-life* 90–98). Acharya Sushruta (*Sushrutasamhita, sutrasthana*, chapter 38) classified plants under thirty-seven sections or *ganas* depending upon their medicinal properties and named them after the most important medicine of that group. For example, the group *rodhradigana* has plants which are antidotal to upset *kapha* (Majumdar, *Plants and Plant-life* 98–104).

Similarly we also find a detailed classification of the plants done by various *acharyas* based on their dietic values. Acharya Charaka (Sutrasthana, chapter 24) recognized six groups of plants based on dietic values and Acharya Sushruta (Sutrasthana, chapter 46) recognized fifteen different groups (Majumdar, *Plants and Plant-life* 105–127). Amara Simha in his *Amarakosha* lists the names of various grains in *Vanaushadhivarga* and *Vaishyavarga*.

Nomenclature

The ancient Indian plant names were based on several parameters. Works like *Raaja Nighantu* and *Dhanvantri Nighantu* have given an outline of these parameters. Here I take up the parameters given in *Rajanighantu* of Pandita Narahari.

Rajanighantu prescribes seven criteria for deriving names:

nāmāni kvacid iha rūḍhitaḥ svabhāvāt deśyoktyā kvacana ca lāñchanopamābhyām /
vīryeṇa kvacid itarāhvayādideśāt dravyāṇāṃ dhruvam iti saptadhoditāni //

(Rajanighntu Grantha-prastavana, 9)

1 Tradition of Usage (*rudhitah*): For example Ocimum sanctum is named as *krishna-tulasi* because it is used in the worship of Krishna and its leaves are blackish green in colour.

2 Properties (*svabhavatah*): For example Embelia ribes is named as *krimighna* which indicates that this plant kills worm and Eclipta prostrata is named as *kesharanjana* or that which blackens the hair.

3 Local names (*deshokti*): Gynandropsis gynandra is named as *hurahura*, which is a local name but it is adopted in the Shastra because of popular usage.

4 Special morphological characteristics (*lanchana*): for example Ricinus communis denoting the palmate leaf of the plant is named as *gandharvahasta* meaning "like the hands of the celestial beings."

5 Similarity or resemblance (*upama*): for example *markatahastatrina* is a beautiful descriptive name indicating the spikes of the plant *Eleusine genus* which resemble the fingers of monkeys.

6 Potency (*virya*): The plant *Plumbago rosea* is named as *agni* because of its potentiality to digest and promote metabolism.

7 Place of growth (*itarahvayadideshat*): *Vanda ruxburghiana* is named as *vriksharuha* because it grows on trees.

Apart from these we find several other parameters enumerated in other texts (A list of texts dealing with various aspects of plant science is given below).

I Names denoting special features

a *Anu*, meaning small or minute, is the name of rapeseed (*raktasarshapa*).
b The name *nyagrodham* for Banyan tree indicates that its shoots hang down.
c *Prasarini* is the name of a creeper that spreads on the ground.
d *Vriksharuha* is the name for a plant which grows on trees.
e *Atmagupta* is the name of *Mucuna pruriens*, which indicates that the fruit of this plant is hidden in a thick cover of stiff and irritating bristles.
f *Shatamuli* is the name for a plant which has many columnar prop roots.

II Names denoting morphological characters

1 Root

a *Bahupadam* is a name of Banyan tree which indicates that it has numerous roots.
b Sweet Flag is named as *shataparvika*. This indicates that the rhizome of this plant has a hundred or many nodes.

2 Knots

a Long Pepper, popularly known as *pippali*, has another name, *granthika*, which indicates that the roots of this plant have a knot-like swollen appearance.

b White birch bark is known as *bhoorjagranthi*, that which has knot- or *granthi*-like fungal formations (*bhoorja*).

3 Bark

a Bamboo is known as *tvaksara* because the stem of bamboo is hollow. There is no heart wood. Hence the name, meaning the bark itself is the core structure.

b The bark tree or *Betula utilis* is named as *valkadrumam* because its bark is used for medicinal purpose.

4 Leaves

a Each leaf of *prasniparni* (Uraria lagopoides) is composed of three leaflets, so the name for this plant is *triparni*.

b Biton Bark, a plant with seven leaflets, is named as *saptaparna*.

5 Flowers

a The flowers of Indian Laburnum (*Aragwadha*) are golden yellow in colour, so this plant is named as *suvarnaka*.

b The flowers of Cobra's Saffron (*Nagakesara*) resemble snakes in appearance, so it is named as *nagapushpa*.

6 Seeds

a The Horseradish (*raktasigru*) is known as *krishnabija* because its seeds of are black in colour.

b Since the seeds of the castor oil plant (*Ricinis communis*) have various marks on them, the plant is named as *chitrabija*.

7 Spines

a The fruits of small caltrops (*Gokshura*) have spine-like projections that resemble the teeth of a dog, so the name given to this plant is *shvadamshtra*.

8 Latex

a   Latex of the roots of Kashmiri Hirtiz (*Euphorbia thomsoniana*) is golden yellow in colour, so it is named as *svarnaksheeri*.

III Names denoting comparisons of plant morphology to other familiar objects

a   Leaves of *Ipomea reniformis* resemble the ears of a rat, so it is named as *akhukarni*.

b   The name *halini* is used for Superb Lily for its rhizomes resemble the plough in appearance.

c   Black Musale is named as *musali* because its tubers resemble the pestle in appearance.

d   Butterfly Pea, popularly known as *aparajita*, has another name, *shankha-pushpi*, because its flowers resemble the conch shell in appearance.

e   *Ricinus communis* is named as *panchangulam* because its leaves are like the palm with five fingers.

f   Musk root is known as *jatamamsi* because its roots resemble matted hair.

IV Names denoting place of origin, habitat, flowering season, etc.

1 Place

a   Cardamom is known as *dravidi* because it is a South Indian crop.

2 Habitat

a   Sacred Lotus is known as *jalaja* because it grows in water.

b   Similarly, *girimallika* which grows in mountains is known as *kutaja*.

3 Season

a   The *bakula* or *Mimusops elengi* is known as *sharadika* because it flowers in the autumn.

b   Swallow Worts or *arka* flowers in all seasons so it is named as *sadapushpa*.

V Names denoting properties of plants

1 Colour

a   Turmeric is known as *pita* because its rhizome is yellow in colour.

2 Odour

a   Sweet Flag (*vacha*) is also named as *ugragandha* because it has a characteristically strong smell.
b   Winter cherry is known as *ashvagandha* because it emits an odour like that of the body odour of a horse.

3 Taste

a   The plant *chiretta* is bitter in taste, so one of its names is *tikta*.

4 Touch

a   The leaves of *Bharngi* (*Clerodendrum serratum*) are rough, so it is named as *kharashaka*.

5 Potency

a   Black pepper is hot in potency, so one of its names is *ushna*.
b   White sandalwood tree is cool in potency, so one of its names is *shita*.

6 Special action

a   *Ipomoea sepiaria* cures sterility and bestows one with progeny so it is known as *putrada*.

VI Names denoting pharmacological actions

a   The plant popularly known as *vidanga* is a good anti-helminthic, so one of its names is *krimighna*, meaning worm killer.
b   Elephant's Foot (*Amorphophallus*) cures rectal haemorrhoids and is known as *arshaghna*.
c   *Aerva lanata* is known as *pashanabhedi* because it has the potency to dissolve stones in urine.

VII Names denoting utility

a   Bastard teak, popularly known as *palasha*, is used in sacrifices, so it is otherwise known as *yajnika*.
b   The wood of *Lagerstroemia parviflora* is used for making chariots, so it is known as *syndanadruma*.
c   Bamboo, popularly known as *vansha*, is used for making bows, so it is also known as *dhanudruma*.

VIII Names denoting important events

a   *Bodhidruma* is the name of the Sacred Fig tree because Buddha became enlightened under this.

IX Names associated with mythology

a   *Guduchi* is also known as *amritasambhava* because it is believed that it originated from ambrosia.

b   *Pinus deodara*, popularly known as *devadaru*, is also known as *indravrksha* because in Indian mythology it is said to be the tree of Indra, the king of celestials.

Thus, we see that the ancient Indian *acharyas* had keen senses of observation which guided them in naming and classifying the plants. We see that though a plant has many names, one of its names is used prominently and all others are its synonyms. So the prominent or most popular name is the *pradhananama* or basionym, and the others are *upanamas* or secondary synonyms. Sometimes we see that one secondary name of a plant is also used as the secondary name of another one or two or three or more plants. On the basis of this, the *nighantukaras* have categorized these names as *ekarthas*, or a name applied to one plant; *dvyarthas*, or one name for two plants; *tryarthas*, or one name for three plants, etc. For detailed information about such names one can see chapters 23–33 of *Rajanighantu* by Pandita Narahari.

Unique set of names for a plant

In this section, I take up the various names of one plant to show what the various names attributed to a plant indicate.

Guduchi or *Tinospora cordifolia* is a climber with long offshoots, rich in foliage and sap, and the leaves of which are used as vegetables. Its mature stem is black green in colour; it has no thorns; it has a bitter taste; it promotes good health and imparts longevity and is benevolent in action; cattle love to eat its leaves; it is capable of rejuvenating itself from the cut bits of the stem; and its fibrous shoot was used in surgery for suturing. This is the description of this plant. Now let us look at the different names of this plant given by the ancient Indian *acharyas*.

The very name *guduchi* is derived from the root "*gud*" which means "to guard or protect or preserve." This indicates the high potentiality of the plant. The names *amritavalli, amritavallari, amritalata, somavalli,* and *somaltika* indicate that this is a weak-stemmed plant. The name *mandali* indicates that the stems of this plant entwines in a circular fashion; *kundali* indicates that the stem gets entangled while it twines; *nagakumari* indicates that the stem has a twining nature comparable to that of young snakes; *tantrika* points out the spreading nature of the plant; *tantri* indicates the tough rope like nature of the plant; *chadmika* refers to its thick foliage; *vatsaadini* indicates that its leaves are eaten by the calves; *shyama* refers to the black-green colour of its stem; *dhaara* indicates that the young stems of this have slight longitudinal grooves; *chakralakshana* indicates the appearance of the stem in cross section; *vishalya* indicates that the plant has no thorny or irritant appendages; the names *china, chinnaruha, chinnodbhava,* and *chinnangi* refer to the undying nature of the stem or stem bits; *abdhikahvaya* refers to the richness of sap in its stem and leaves; *amrita* indicates that the persons using this plant would live a long and healthy life; *soma* refers to the powerful action of the plant as an elixir; the names *rasayani, vayastha,* and *jivanti* refer to the rejuvenating nature of the plant; *jvarashini* and *jvarari* refer to the specific use of the plant in fevers; *bhishakpriya* and *bhishakjita* signify that this plant is the favorite of the physicians; *vara* indicates that it is the best among medicines; *soumya, chandrahasa,* and *chandahasa* indicate its nature of benevolence in action; and *devanirmita, amritasambhava,* and *surakrita* indicate the divine origin of the plant.

Though this way of naming plants is considered to be superior to the modern system, it also has its own limitations and disadvantages. The widespread use of synonyms (*paryaya*) and homonyms (*nanartha*) for virtually every plant has resulted in great confusion. This is confounded by a lack of authentic preserved material and of descriptions of the plants concerned. We also notice the depletion of traditional expertise in the identification of various plants. We see in the classical treatises dealing with the classification of plants that the principal names, or the basionyms, have not been strictly adhered to or pointed out. So in most of the cases it is not possible to know which is the basionym of the plant. There also has been difference of opinion regarding the identification of plants among the commentators. For example, the plant called *jingini* (*Dialium coromandelicum*), or *jiyol* as it is called in West Bengal, is identified as *krishnatulasi* (*Ocimum tenuiflorum* or Holy Basil) by Indu in his *Indunighantu*; as *manjishtha* (*Rubia cordifolia* or Indian madder) by Chakrapanidatta in his *dravyagunasangraha*; and as *shankhapushpi* (*Convolvulus pluricaulis*) or *aparajita*, as it is popularly known in the *ashtangahridayakosha*, a compilation of technical terms used in the *ashtangahridaya* of Vagbhata as prepared by Velapada K. M. Vaidya. This has led to utter confusion. Sometimes the lexicographers have used wrong synonyms. There are also copying mistakes in the manuscripts available which lead to controversy. The use of substitutes also has led to the problem of identifying the original plants. However, a detailed and comparative study of ancient medicinal as well as non-medicinal treatises dealing with plant sciences can enable us to understand better not only the principles but also the advantages and disadvantages of the ancient Indian system of nomenclature and classification of plants.

Conclusion

The ancient Indian *rishis* believed that each and every element in this creation has a consciousness. There is nothing without consciousness. With this understanding, when the ancient Indian botanical scientists dealt with the plants they could feel the consciousness of the plants. As Manu says,

> *tamasaa bahurupena veshtitaa karmahetunaa.*
> *antahsamjnaa bhavantyete sukhaduhkhasamanvitaah.*
>
> (Manusmriti, 1.47)

Though it seems that the plants are covered by *tamas*, they are conscious within, and because of this they feel the pleasure and pain. We also see it in the *Yogavasishtha* (6.1.10.23), where Sage Vasishtha makes it clear that the consciousness power (*citshakti*) lies dormant in all that is inanimate (*citshaktirvaasanaabijarupini svaapadharmini sthitaa rasatayaa nityam sthavaraadishu vastushu*). Udayana in his book *kiranavali* under the section *prithvinirupana* speaks about the phenomena of life, death, sleep, disease, and movements in the plants. Gunaratna in his commentary on *shaddarshanasamuccaya* also deals with the plant consciousness and gives a list of plants that display the phenomena of sleeping and waking. The Buddhist logician Dharmottara in his *Nyayavindutika* records the phenomenon of sleep in certain plants in the form of contraction of their leaves during night. Sankaramisra, in his *Upaskara*, talks about *bhagnaksatasamrohana* of plants, which means there is a natural healing after injury due to the growth of organs. The *Shantiparva (184.10–17)* of *Mahabharata* enumerates several physiological principles, including the sense of touch, hearing (response to sound), vision, smell, irritability, etc. in respect of plants.

घनानामपि वृक्षाणामाकाशोऽस्ति न संशयः।
तेषां पुष्पफलव्यक्तिर्नित्यं समुपपद्यते॥

उष्मतो म्लायते पर्णं त्वक् फलं पुष्पमेव च।
म्लायते शीर्यते चापि स्पर्शस्तेनात्र विद्यते॥

वाय्वग्निन्यशनिनिर्घोषैः फलं पुष्पं विशीर्यते।
श्रोत्रेण गृह्यते शब्दस्तस्माच्छृण्वन्ति पादपाः॥

वल्ली वेष्ट्यते वृक्षं सर्वतश्चैव गच्छति।
न ह्यदृष्टेश्च मार्गोऽस्ति तस्मात् पश्यन्ति पादपाः॥

पुण्यापुण्यैस्तथा गन्धैर्धूपैश्च विविधैरपि।
अरोगाः पुष्पिताः सन्ति तस्माज्जिघ्रन्ति पादपाः॥

पादैः सलिलपानाच्च व्याधीनामपि दर्शनात्।
व्याधिप्रतिक्रियत्वाच्च विद्यते रसनं द्रुमे॥

वक्त्रेणोप्पलनालेन यथोर्ध्वं जलमाददेत्।
तथा पवनसंयुक्तः पादैः पिबति पादपः॥

सुखदुःखयोश्च ग्रहणाच्छिन्नस्य च विरोहणात्।
जीवं पश्यामि वृक्षाणाम् अचैतन्यं न विद्यते॥

In the modern time scientists like J. C. Bose and others also confirmed the truthfulness of the above facts. To find out whether the plant perceives and reacts to a shock, the stem is placed between a fixed rod and a movable magnifying lever, the movement of which is further magnified by optical means to something like a million times. The experiment described was carried out before a group of world-famous scientists of the University of Vienna. A very feeble electric shock was sent through the plant and through one of the leading physicians who was in the same circuit. The human being felt nothing, but the Contraction Recorder showed that the plant gave a shouldering twist under the shock (*J. C. Bose Speaks*, 1985, Calcutta, 355).

Sri Aurobindo says:

> Chit, the divine Consciousness, is not our mental self-awareness; that we shall find to be only a form, a lower and limited mode or movement. As we progress and awaken to the soul in us and things, we shall realise that there is a consciousness also in the plant, in the metal, in the atom, in electricity, in everything that belongs to physical nature; we shall find even that it is not really in all respects a limited mode than the mental, on the contrary it is in many "inanimate" forms more intense, rapid, poignant, though less evolved towards the surface.
>
> (Complete Works of Sri Aurobindo Volume: 23–24,
> The *Synthesis of Yoga*, 387)

It was with this attitude that the ancient Indian medical scientists and botanists dealt with plants. Although *pariksha*, or investigation, was one of the prominent means for them to validate their knowledge, they relied mostly on their intuitive mind. Discovering by intuition and developing by logic was their method in systematizing knowledge. If we see how the Mother of Sri Aurobindo Ashram gave spiritual names to hundreds of flowers, if will help us understand the ancient Indian methodology and philosophy of naming and classifying the plants. Once one of the Ashramites asked the Mother, "Mother, when flowers are brought to you, how do you give them a significance?" Then the Mother replied by saying, "by entering into the contact with the nature of the flower, its inner truth. Then one knows what it represents" (*Complete Works of the Mother*, Vol.5, P. 230). It was with this spirit that the ancient Indian *acharyas* gave names to everything by entering into the soul and nature of everything. The thousands of names of plants available in various ancient treatises are the expression of this inner contact with the nature and soul of the plants.

Works cited

Primary sources

Amarasimha. *Amarakosha*. Ed. Pandit Haragovinda Sastri. Varanasi: Chaukhamba Sanskrit Sansthan, 1982.

a. *Krishiparasara* of Parasara, Ed. and Trans. Girija Prasanna Majumdar and Suresh Chandra Banerji. Kolkata: Asiatic Society, 1960.

b. *Krishi-Parashara* (Agriculture by Parashara) Trans. Nalini Sadhale. Secunderabad: Asian Agri-History Foundation, 1999.

Hemachandra. *Nighantushesha*. 12th cent. Ahmedabad: L. D. Institue of Indology, 1968.

Kashyapa. *Krishisukti*, Ed. and Trans. Gyula Wojtilla. Budapest, Hungary: Acta Orientalia Academiae Scientiarum, 1985.

Kayyadeva. *Kayyadevanighantu or Pathyapathya-vibhodhaka (1450AD)*. Ed. P.V. Sharma and Guruprasad Sharma. Varanasi: Chaukhamba Orientalia, 1977.

Madhava. *Paryayaratnamala*. 9th cent. Ed. Tarapada Chaudhury, *Patna University Journal*. Vol.2 (1946).

———. *Madhavadravyaguna or Bhavasvabhavada*. 13th cent. Ed. P.V. Sharma. Varanasi: Chaukhamba Vidya Bhavan, 1973.

———. *Madanapalanighantu or Madanavinoda or Madanapala (1374AD)*. Ed. P.V. Sharma. Varanasi: Chaukhamba Sanskrit Series, 1977.

Mallinatha. *Abhdhanaratnamala*.Ed. P.V. Sharma. Varanasi: Chaukhamba Orientalia, 1977.

———. *Raaja Nighantu and Dhanvantri Nighantu*. Ed. Vaidya Narayana Sharma. Pune: Anandashram Press, 1986.

Ravigupta. *Siddhasaranighantu*. 9th cent. Ed. R.E Imeric. Weisbaden: Fraz Steiner Verlag Gmbh, 1980.

Sarangadhara. *Upavana Vinoda*. Trans. G.P. Majumdar. Kolkata: Satis Chandra Seal, The Indian Research Institute, 1935.

Shodhala. *Shodhalanighantu*. 12th cent of Gaekoward Oriental Series. Ed. P.V. Sharma. Baroda: Baroda University, 1978.

Sri Bhavamisra. *Bhavaprakasa Nighantu (Indian Materia Medica)*. Ed. G.S. Pandey and K.C. Chunekar. Varanasi: Chaukhamba Bharati Academy, 1999.

Surapala. *Vrikshayurveda [The Science of Plant Life]*. Ed. and Trans. Nalini Sadhale. Secunderabad: Asian Agri-History Foundation, 1996.

Vahata. *Ashtanganighantu*. 7th to 9th cent. Ed. P.V. Sharma. Chennai: Kuppuswami Sastri Research Institute, 1973.

Varahamihira. *Brihat Samhitha*. Trans. M. Ramakrishna Bhat. New Delhi: Motilal Banarasidass, 1992.

Secondary sources

Bibliotica Indica; Royal Asiatic Society of Bengal; Collection of Oriental works, Calcutta Baptist Mission Press, New Series No. 1342 (1917): 256.

Majumdar, Girija Prasanna. *Plants and Plant-life as in Indian Treatises and Traditions*. Kolkata: University of Calcutta, 1927.

Manohar, P. Ram. "Nomenclature and Taxonomy." *Vrikshayurveda*. Chennai: Centre for Indian Knowledge System, 1994.

Sivarajan, V.V. *Introduction to the Principles of Plant Taxonomy*. Ed. N.K.B. Robson. 2nd ed. Cambridge: Cambridge University Press, 1991.

The Vizianagram Sanskrit Series, Vol. IV. Benaras, 1895.

Other texts

Ashtangahridaya of Vagbhata
Behlasamhita
Charakasamhita of Charaka
DakshinamurtiNighantu by Venkateswara
Dravyagunashatakam or shatashloki of Tirumalla Bhata
Haritasamhita
Hridayadipakanighantu of Bopadeva
Indunighantu of Indu
Kashyapasamhita
Nighanturatnakara of Vishnudeva
Nighantusangaraham of Raghunathji
Paryayarnava by Nilakantha Mishra
Rajavallabhanighantu of Rajavallabha
Shabdapradipa by Surapala
Shaligramanighantu of Shaligrama
Shivakosha of Shivadatta Mishra
Shukraniti of Shukracharya
Siddhamantra of Keshava
Sushenavaidyaka or Ayurveda Mahonidhi
Sushrutasamhita of Sushruta
Vaidyavatamsham of Lolimbaraja

12

A MACRO-MICRO SYSTEMS APPROACH TO FRAME HOLISTIC STUDIES ON THE CULTURE AND PHILOSOPHY OF SCIENCE IN INDIA

Joy Sen

Proposals for holistic frameworks for conducting studies on culture and philosophy of science in India can be made comprehensive by integrating two dimensions:

A. Experiments that are particular level-based and
B. Realizations that are based on iterative hierarchy-based assessments

The two are often accepted as conflicting by the classical scientific world. But there is a new and emerging scientific temper that increasingly recognizes the strength of complementarities that lies deeper between the two. Conventional scientific temper, as derived from classical Western schools, is a product of particular level-based assessment only. It is usually reductionist, Cartesian, and linear in approach. The approach is limited to a linear cause-and-effect correlation, modelled at a predominantly material plane and disjointed to serve only the modernization-industrialization paradigm. The approach broadly characterizes much of our current notions of global growth and development.

However, since the assertion of relativism even in the sciences, a new recognition or temper has originated. The new temper and its inherent paradigm press for a more non-linear, non-reductionist approach to scientific model making. It is not isolated from the totality covering reality and humanity, and is also not divorced from its entire lifeline embracing both cultural studies and philosophy. On the one hand, this temper is born out of an evolutionary process of humankind. On the other hand, it emerges from a growing feeling of uncertainty in every sphere of current materialist-reductionist knowledge and in the global economy, and also from improper distribution of wealth and access to ecology, climate, health, education, and equitable opportunities all around the world. India is a part of this uncertainty. So is scientific pursuit in India, which is indeed a sub-system. A lopsided, ethnically and racially split, "developed/underdeveloped" view of development has fallen short of achieving a healthy humane world. Therefore, this new paradigm questions the classical scientific standpoint on human development and demands urgent shifts and changes.

This chapter addresses such grave concerns. It addresses the need to redraw complementarities between scientific pursuits at a particular level of knowledge with pursuits at other levels, which may not be strictly "science" in the reductionist sense, but perhaps a larger science with its wider, comprehensive framework. The other levels are culture, philosophy, and normative sciences like psychology and bio-anthropology. They may extend up to "deep ecological" or "spiritual" dimensions that empower the framework to be holistic.

The discussion is carried out in two parts. Part one briefly explains the background of original wisdom of holism, which had existed in the ancient world. Then it explains a departure from that wisdom to phases of manipulated contraction, guided limitation, and consequential reductionism. Part two subsequently forwards five major examples explaining a return – a turnaround of the present state of reductionism to holism – to that ancient wisdom itself.[1]

Part one

From holism to disjointed-incrementalism: origin & departure

Holistic frameworks integrate knowledge systems between levels in a hierarchy. Integrations therefore form composite wholes as opposed to parts or reduced identities. The whole represents interpenetrations between relatively smaller micro levels – like science, culture, and philosophy – and the macro setup, which may extend from continental spaces to very large backdrops like humanity itself in relationship to the cosmos. These interpenetrations had been the basis of Indian spirituality – the wisdom of the sages and their realized experiments with truth, with nature, and with humanity. The realization (*vid*) of the impersonal and universal truth, independent of time and any personal space, is called the Vedas.

The earliest Greek scientific world of the Milesians believed in such integral approaches. It is only from the Eleatic school that departures and manipulations occurred in the West. It is evident from the words of Fritjof Capra:

> the monistic and organic view of the Milesians was very close to that of ancient Indian and Chinese philosophy, and the parallels to eastern thought are even stronger in the philosophy of Heraclitus of Ephesus, Pythagoras of Samos.

The split of this unity began with the Eleatic school, which assumed a divine principle standing above all gods and men. This principle was first identified with the unity of the universe, but was later seen as an intelligent and 'personal" god who stands above the world and directs it. Thus began a trend of thought which led, ultimately, to the separation of spirit and matter and to a dualism which has become characteristic of Western philosophy."

(Capra 5)

During a key conversation between Plato and Aristotle, the two harbingers of Greek philosophy and sciences, Plato is said to have requested the individual aspiration to look up, evolve as an individual from self-centeredness to a universe-centered human being. On the contrary, Aristotle said that the individual has to come down to reality, be practical, and allow the individual only to take over and rule. Individual freedom has to overrule (unlike Plato at the universal plane) at the predominant material plane.

Nearly two and a half thousand years ago Plato and Aristotle had a momentous disagreement about communes and the commons. Plato believed that the family, private property, and ownership in general foster clannish self-interest; these alternative arrangements are intended to do the opposite and promote "a universal feeling of sympathy" in which the greater good of the whole community takes precedence over individual whim.

Aristotle, with what seems like unanswerable reasonableness, dismissed Plato's ideas as simply impractical: "That which is common to the greatest number," he writes in *Politics*, "has the least care bestowed upon it. Everyone thinks chiefly of his own, hardly at all of the common interest; and only when he himself is concerned as an individual." In a striking phrase, Aristotle suggests in the text that in Plato's ideal state, "love will be watery. . . . How much better it is to be a real cousin of somebody than a son after Plato's fashion!"

Like so many of today's mainstream economists and politicians, Aristotle defends the claims of private property and family values: people, he argues, can only really love and care for what they own or have an intimately personal stake in. Aristotle would not be mistaken for a liberal, but he nevertheless questions some of the assumptions inherent in the Republic. Most importantly, Aristotle's *Politics* is a more pluralist understanding of government because Aristotle argues that citizens with proper education and obedience to the law are equipped to rule themselves and others. These citizens also come together to rule in the multitude – or majority – something that scares Plato and epitomizes mob rule. "[M]any and not one should rule," suggests Aristotle, "because anyone can rule well when educated by the laws and many ruling together and better than one ruling alone" (Aristotle 110).

Aristotle's opinion of virtue is similar and stands in opposition to Plato. While Plato believes that the virtuous are small in number and that their

137

virtue is inherent, Aristotle thinks that virtue and justice can be taught to citizens (*Resurgence* Issue 235). Thus a gap between exclusive intuition, which is person-specific, and mass education. trying the offer the same "block" to everyone, emerged. Much of the modern education system is based on this mass "block" packaging. The issue of this "packaging" of education lies at the heart of modern societal crisis – as pointed out by current Indian thinkers like Swami Vivekananda, Sri Aurobindo, Rabindranath Tagore, M. K. Gandhi, and, of late, S. Radhakrishnan and many others.

Plato was worried about mob rule and suspicious of any idea that gave ruling power to common citizens. Plato also resisted change and charged the guardians with being responsible for seeing that the laws and arts were protected. As well, Plato believed that "national" myths, noble lies, and religion were important, if but for no other reason than to provide order. Aristotle, Plato's counterpart, disagrees on these issues. Aristotle is more willing to give citizens the ability to rule, is not as resistant to change, and ultimately thinks that with the proper laws and teaching the citizens would have the ability to govern themselves. Though Plato and Aristotle are our "ancient" writers, their disputes and disagreements are still apropos to today's political dialogue (*Resurgence* issue 235).

The holistic "Platonic" view of the world was opposed to that of the Aristotelians, who forerun the later-day European philosophy and its mass education system. Though Plato and Aristotle are the apparent personification of a significant disagreement, there lies deeper an impersonal cause – strife between the two ideas of time. One, where time is holistic, i.e., long-ranged, aiming for the overall order. In such a view of time, smaller temporal forces of lesser good (evil) and greater good (good) are reconciled in larger wave "patterns of changes." The other view of time is momentary and short-ranged, concerned only about things right now and therefore aggressive and differing to patient linkages between "humanity" and "nature."

The Aristotelian approach, and more so that of the Eleatic school, accommodated the short-range viewpoint, sowing the seed of mainstream European reductionism. An absoluteness of material functionalism and technology became the guiding principles.

An integrated "Man-Nature dynamics" of the ancient Greeks – preceding Aristotle and found in Plato, Pythagoras, and Protogoras – was lost as the Vitruvian school of Rome gradually succumbed to later changes . . . to the forces of a machine driven Industrial European civilization (Sen 2007).

Later Western philosophy of science had propagated a "Machine Age" of divided man-nature dynamics; it has shaped much of the mercantile imperial age of the last century and, of late, the corporate managerial dictums of a so called "globalized" world. In the words of Fritjof Capra, it reflected the impact of Semitic religious wisdom that complemented the divided Cartesian machine order of the West:

The view of man as dominating nature and woman, and the belief in the superior role of the rational mind, have been supported and encouraged by the Judeo-Christian tradition, which adheres to the image of a male god, personification of supreme reason and source of ultimate power, who rules the world from above by imposing his divine law on it. The laws of nature searched for by the scientists were seen as reflections of this divine law, originating in the mind of god.

("The Turning Point" 24)

Crisis of reductionism

For the last three centuries, with the rise of the "Machine Age," scientists have used "the reductionism's approach" more and more. They have based their ideas on the Newtonian approach and the scientific philosophy advocated by Sir Francis Bacon. The approach, still the ruler of our own times, perceives matter to be the basis of all existence and conceives the material world as a multitude of "disjointedness" – of separate objects assembled into a huge machine. This is the microcosm.

The macrocosm is left as disjointed and conceived as a similar machine, only larger in size. Over time and development of the eighteenth-century sciences, it was believed that all complex phenomena could always be understood by reducing the various phenomena into "building blocks" or packages and analyzing the mechanism through which they interacted. This is the spirit of reductionism.

A dividing line between one phenomenon and the other phenomenon was drawn, destroying or subverting the very network that integrates human beings (behavior or the microcosmic pattern) and nature (environment or the macrocosmic pattern). The dividing line forwarded a one-legged explanation of human growth and development (evolution, purpose, and functions) only, and it was also on the basis of competition and/or best adaptation by disintegrating or splitting an integrated dynamism of inner human growth and the growth of nature. The chief architects of this disintegration were René Descartes, Charles Darwin, and Sigmund Freud.

Descartes had been a major driver and founder of reductionism. Descartes' entire view of nature is based on the fundamental division between two important and separate realms called the "mind" (res cognitas), which is the thinking being, and "matter" (res extensa), the extended being. To him, the material universe is a machine with no purpose, life and spirituality in matter sharply dividing an abiotic material realm from the "biotic" mind. Nature worked according to the Cartesian division of mechanical laws, and everything could be explained in terms of the arrangement and movements of its parts. This mechanistic or reductionism's viewpoint of reality guided all scientific observations and shaped its very foundations. This led to a drastic change in the image of nature from organism to a machine, and this affected people's attitude towards the natural environment – an effect that has accumulated to the building up of an inorganic machine style of built-environmental design at the expense of natural resources and passive-design capabilities (Merchant).

Charles Darwin's theory of evolution was founded on the twin concepts of "chance variation" and "natural selection." It soon became apparent that chance variations, as conceived by Darwin, could not explain the emergence of new characteristics (like creativity) in the evolution of species. The solution to this greatest problem of the Darwinian viewpoint was discovered by Gregor Mendel and that gave way to the "reductionism" put forward by the "genetic determinism or stability through adaptation" of the Darwinian approach. Soon, the co-existence of the other dimension – "mutability" (the ability to transform, change, create new changes) – evolved.

The classical works of Sigmund Freud were limited to inferences drawn from experiments on a particular class of people conforming to retarded and animal-like instincts. From here, the three later works of Carl Gustav Jung,[2] Abraham Maslow,[3] and lately, Ken Wilber[4] have come a long way. All three of them have gone beyond the limited range of human functions and intelligence which was restricted to "competitive animalism" and "a survival instinct of the fittest." The critical damage imparted by the architects of reductionism is serious and is a subject of cross-examination and re-evaluation. It is evident from the following words:

> In investigating the roots of our current environmental dilemma and its connections to science, technology and the economy, we must re-examine the formation of a world-view and a science which, by re-conceptualizing reality as a machine rather than a living organism, sanctioned the domination of both man and woman. The contributions of such founding "fathers" of modern science as Francis Bacon, William Harvey, René Descartes, Thomas Hobbes and Isaac Newton must be re-evaluated.
>
> (Merchant, "The Death of Nature" xvii)[5]

Part two

The Indian perspective and its contribution to the new temper

In Indian philosophy, culture, and religions, holistic approaches are not new. Integral approaches to the science of cosmology and culture based on nature had started with the very ancient Upanishadic foundations and allied schools initiated earlier by Vedic seers in the ashrams that were nature-based, resource-conscious, green, and symbiotic. Sages Yuggyabalka,

Agastya, Kapila, and Dattatreya are to name only a few. This continued till the times of green *viharas* of Buddha and Mahavira.

The integral visions of Vedic ages have been sustained by many interim approaches as propounded by the Natha School, the Nimbarka School, and the Advaita School of Kashmir. Such approaches are even continuous today – right up to the times of Sri Ramakrishna and Sri Aurobindo in our own days, which all have been well received in the West in the last few decades. This has been true particularly after the ravages of the two great World Wars, the great economic depressions of 1929 and 2009, and the climatic-ecological crisis that is faced by the majority now.

The aim of this part therefore is to put forward five major examples that carry in themselves the evidence of contributions by Indian wisdom towards the making of the new scientific temper of holism in the West. It is in these five examples the spirit of the revival of holistic temper can be best felt. This temper has the innate capacity to render true symbiosis between science, philosophy, and culture – for India and for all humanity.

Forwarding a new temper: five major examples

Example one: bio-cosmology and complementarities

The historic interaction between India's patriotic monk-reformer Swami Vivekananda and electrical scientist Nikola Tesla took place after the Parliament of World Religions in Chicago in 1893. The basis of these interactions inspired key areas of bio-telecommunication and genetic algorithms in the days to come through the works of Tesla, thereby holding the promise of integrating the planetary existence on a higher platform of a world wide web that already existed, at least in a crude form.

After the close of the Parliament on 27 September 1893, there was an invitation to the house of Elisha Gray, the noted inventor of electrical equipment (Life, Vol. 1: 448). Later around 28 and 29 September 1893, the Swami had gone to Highland Park.[6] Some of the important guests at Highland Park were the British physicist Sir William Thomson (also known as Lord Kelvin); the German scientist Herman Von Helmholtz; and the French delegate to the International Electrical Congress, Professor Edourd Hospitalier. Here the name of Nikola Tesla was never heard, at least officially. But some researchers suppose that he was there (Chowdhuri 127). Tesla was the inventor of the alternating current[7] generating machine, which was being demonstrated at the Chicago World Exposition. It is possible that Tesla had met the Swami at Chicago during and after the Exposition, as both of them were centers of attraction. But real interactions happened later, and these were in New York.

The proper mention of Tesla in Vivekananda's letters[8] comes later, on 13 February 1896. In this letter, Swamiji emphasizes his conversation with Tesla on an area that was unique. Surprisingly, the meeting with Tesla was in an auditorium where Swamiji had gone to see the play *Iziel* – acted by actress Madame Bernhardt. Incidentally, *Iziel* happened to be a play depicting a discourse between Lord Buddha and a dancer-girl named Iziel. Swamiji had first talked to Madame Bernhardt and then to the famous singer Madame Morelle – who were the masterminds behind the theatrical production. Finally he interacted with Tesla.

Swamiji explained cosmology on the basis of three variables. First, *Prana*, a life current constituting all dynamism from small to large; second, *Akasha*, an all pervading space from small to large; and third and the last, *Kalpa*, the single cycle of time representing a unitary cosmic creation and dissolution. Recalling Tesla's reaction to the invocation, Vivekananda writes:

> Our friend (Tesla) was charmed to hear about the Vedantic Prana and Akasha and the Kalpas, which, according to him, are the only theories modern science can entertain. Now both Akasha and Prana again, are produced from the Cosmic Mahat, the universal mind, the Brahma or Iswara. He (Tesla) thinks he can demonstrate mathematically that force and matter are reducible to potential energy. I am to go and see him next week, to get this new mathematical demonstration.
>
> (Swami Vivekananda, Letter to Mr. Sturdy)

A brief summary of the complementarities between the two apparent conflicting worlds is explained in Figure 12.1. Here cosmology is best explained kinetically on the interactive twin or conjugate of microcosmic life and space (*prana* and *akasha*), and its higher potential realized as macrocosmic totality or *Mahat*.

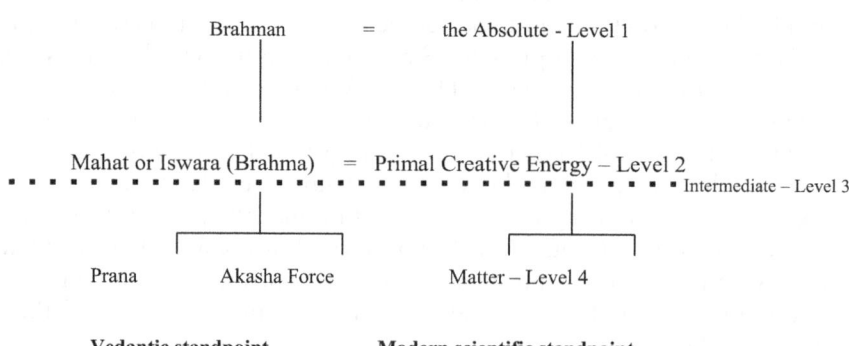

Figure 12.1 Parallels of Indian wisdom to Western sciences

Example two: bio-physics – basis of anthropic foundation

Later in 1930, poet Rabindranath Tagore and scientist Albert Einstein met, and their conversation laid the foundations of another aspect – that of a participatory application in sciences called the "anthropic foundation" – explaining an involvement of human consciousness in mirroring universal experience and empiricism. Later scientists like Robert Penrose and Stephen Hawking and many others continued along these pointers. Gradually, there emerged a field of binary brain research called neuroplasticity that explains the mirroring of the individual and cosmic worlds based on binary qualities like analyses and intuitive thoughts.

In 1930, poet Rabindranath Tagore and physicist Albert Einstein had a discussion, which is closest to the anthropic principle. Here is a part of that historic dialogue.

- Einstein: There are two different conceptions about the nature of the Universe: a) The world as a unity dependent on humanity; and b) The world as a reality[9] independent of the human factor.
- Tagore: When our universe is in harmony with Man, the eternal,[10] we know it as truth and we feel it as beauty.
- Einstein: This is purely a human conception of the universe.
- Tagore: There can be no other conception. This world is a human world – the scientific view of it is also that of the scientific man. There is some standard reason and enjoyment, which gives it truth, the standard of the eternal man whose experiences are through[11] our experiences.
- Einstein: Truth, then, or Beauty, is not independent of Man?
- Tagore: No.
- Einstein: If there would be no human beings any longer, the Apollo of Belvedere would no longer be beautiful?
- Tagore: No . . . there is the reality of paper, infinitely different from the reality of literature. For the kind of mind possessed by the Moth, which eats that paper, literature is absolutely non-existent, yet for Man's mind literature has a greater value of truth than the paper itself. In a similar manner, if there be some truth which has no sensuous or rational relation to human mind it will even remain as nothing so long as we remain human beings.

Finally two great minds were in perfect agreement on a worldview as a unity of involved external nature (macrocosm) and dependent on the evolution of human perception (microcosm).

Example three: bio-cosmology – evolution of non-locality
and grand unification

Next, in the late World War II years, after the Manhattan Project and Princeton years, the foundations of larger implications of bio-cosmology and bio-anthropic foundations were initiated. Works of Satyendra Nath Bose, in alliance with that of Einstein and later that of Werner Heisenberg and Neils Bohr, had inspired a whole new generation of people – from John Bell to scientist Alain Aspect in Europe. This has been the beginning of a new theme in physics called teleportation, seeking an event horizon and integrity of information transfer in the universal reality, which can be only explained by realms beyond the speed of light. Eventually, its linkages with quantum electrodynamics and participatory human consciousness were established to a certain extent.

A sequence of changes in modern physics was on their way since the days of Albert Einstein – all leading to the theory of teleportation or non-locality (Sen, *Concept of Complete Religion*):

- In May 1935, Albert Einstein and two of his colleagues, Boris Podolsky and Nathan Rosen, published the first description of a reality that is non-local (meaning traveling faster than the speed of light) as something unbelievable. Einstein called this "spooky action," and this publication was eventually called the EPR paradox. It includes two major enigmas: (1) the collapse of the wave function as interpreted through the wave equation of Erwin Schrödinger and (2) a probabilistic notion that every part of a quantum system responds instantaneously to a stimulus affecting any other part of a system. They suggest instantaneous connection with the universe as a whole.

- The EPR experiment inspired other areas of research like polarization of light[12] seen from a quantum viewpoint. These emissions suggested information transfer in opposite directions to points far apart. Whether such transfer actually takes place at a speed exceeding that of light is another question.

- Two later works emerged. One was of David Bohm (London, UK) in the 1950s, and the other was by John Bell at CERN (European Center for Particle and Nuclear Physics, Switzerland) in the 1960s. Bohm proposed the "Scattered-Matrix" theory and John Bell his "Bell's Theorem" to take the notion of "non-locality" further. But all these were theories.

- In 1980s, Alain Aspect and his colleagues (in Paris) for the first time experimentally confirmed the EPR enigma and posed new avenues to review the Copenhagen interpretation of Schrödinger.

- In the early 1990s, John Cramer at the University of Washington, Seattle, pointed out a unique thing implied in Schrödinger's wave equation. Cramer reviewed the equation and explained a bi-polarity meaning a conjugate of opposites. His wave equation is based on the simple idea of complex numbers (Gribbin 1995)[13] in any quantum calculation or experiments. Cramer's deliberations imply that a wave function has a bi-polarity of backward and forward linkages of time. Backward linkage of time takes creation backward to its roots, its origin. In other words, the linkage is one of anti-creation (deconstruction taking everything to its seed form). Forward linkages of time, on the contrary, take us to future. Thus the Indian concept of *Tri-kala* is integrated well within Western sciences.

Example four: bio-ecology – evolution of chaos theory and overview effect

Integration of global-local systems combining bio-physics and ecology had begun with many – the names of J. C. Bose in India and E. F. Schumacher, Eric Jantsch, and Ilya Prigogine in the West are only a few among them. On the whole, the spirit of integration continued to inspire people from Rachel Carson to Al Gore. Births of new fields like ecological economics and nano-biosciences have paved a new way to integrate large-scale data structures within an earlier reductionist model in chemistry, physics, and eco-biology. This leads to nano-sciences.

Neurologist Robert Livingston suggests that the evolution of life is a selection process acting on the basis of behavior. Systems theorist Erich Jantsch puts it further that, in the history of human living, the co-evolution of macrocosm and macrocosm is of great importance. Jantsch says that conventional accounts of the origin of life usually describe the building of higher life forms in microevolution and neglect the macroevolution. But the two are complementary and part of the same evolutionary process, as he points out. This is what chemist James Lovelock and microbiologist Lynn Margulis mean by the "mind of Goddess Gaia" – where individual minds are embedded in the larger minds of social and ecological processes, participating further in the universal mind of the cosmos – symbolized by "Gaia's mind."

The crisis of uncertainty in the reductionist paradigm and the need of a larger framework are evident in the statement made by the 1977 Nobel Laureate in Chemistry:

> Classical science has emphasized stability and equilibrium. We see today a different world – the world of fluctuation, of evolution, of instabilities and we have to understand where this comes from. Essentially I will tell you about "Chaos" to show you that with "Chaos" we enter a new physics and mathematics, which may, in my opinion, lead to revolution in science comparable to the ones due to quantum mechanics and relativity.
>
> I like to show you that the idea of physics is in some sense changing. The classical idea of that is a kind of geometry in which "Time" does not play an important role. Certainly there is no difference between future and past. But, of course, the problem of "Time" is of many philosophical aspects.

145

Every great civilization like Indian civilization has speculated about the nature of "Time." Is the universe ruled by deterministic laws? There is no conflict between being and becoming. It is the central problem studied in the Vedas – this problem of overcoming the position of being and becoming. I am not an expert on the Vedas. But certainly when you look at the cosmic dance of Shiva or the dream of Vishnu you see how to overcome the aspects of being to becoming.

(Ilya Prigogine. Nobel Laureate. Chemistry, 1977)

Example five: ecological-psychology – recognition of a truth integrating nature and the human through transpersonal psychology and deep ecology

Contributions of the realized ideas and works of Sri Aurobindo, and cosmologists like Pierre Teilhard de Chardin have inspired a whole new generation approach to individual psychology having larger implications of altruism and philanthropy. The works of C. G. Jung, Abraham Maslow, Ken Wilber, Andrew Cohen, and many others of the Trans-personal schools, the Systems schools and the Deep Ecological schools are living examples – all exhibiting a new disposition to integral sciences – sciences that are both humane and cosmic.

Carl Gustav Jung proposed a theory of psychological types by identifying four characteristic functions or levels of human psychology. They are (1) the level of sensation at the gross body level – the very level on which Freud mostly concentrated; (2) the level of thinking and intelligence from where mainly scientific and rational functions emerge; (3) the level of feeling and aesthetics that facilitates collective causes and value systems of larger humanity; and (4) the level of intuition that goes further to even larger ecological and natural understandings of human anthropology in relation to overall nature and the final purpose of human evolution.

Abraham Maslow's "theory of trans-personation and self-actualization" followed Jung's view. Maslow had rejected Freud's view of human intelligence dominated by lower instincts or animalism. The reason Maslow gave for this rejection was Freud's experimental and consequential inferential limitation based on the study of neurotic and psychotic individuals coming from a strained part of the society which was afflicted by the perils of earliest industrial revolutions. Maslow and his colleagues stressed that human personality and intelligence could not be separated from the network of human relations in which it exists. He defined psychology as a discipline dedicated to the study of interpersonal relations and interactions – the key to social intelligence. From here, transpersonal psychology begins to recognize, understand, and ultimately realize the "wider or beyond the single individual" states of consciousness and consequential intelligent networks.

The recent work of Ken Wilber, called "spectrum psychology," further contributes to the evolutionary journey from personal to inter-personal and from the inter-personal to the trans-personal states. Wilber's approach cohesively assimilates numerous models, both Eastern and Western, and presents a hierarchy of progressive bands or "spectrums" of consciousnesses like the seven colours of a rainbow, like a septuplet matrix. Each band or level is characterized by a different state of identity, ranging from supreme or cosmic consciousness and intelligence as a primal unity to downward levels of the narrow identity of the ego-intelligence, which we may term the weaker individual Anthropic principle.

Such re-evaluations have begun, and they are coming from all sectors of knowledge – physics, biology and bio-sciences, psychology, medicine, economics, environmental sciences, ecology, and all forms of art. In all of them there is an attempt to include what was lost and consequentially link the "reduced" to the "whole."

Conclusion

Perhaps humanity is about to enter a phase of an integrated order of social and engineering systems based on advances of human communication and new areas of "human-nature dynamics." The principle of life (Bios) and its integration with equitable human-resources development are increasingly seen as the crying need of resource-conscious, planetary-conscious, cosmos-conscious social and ecological ethics. The vision is expected to improve and expand further in the days and years to come. Only the initial progresses have been made evident by the discussion in part two.

At the same time, people have begun to realize the increasing hazards of a mechanistic worldview, and that is pressing current sciences and technology to look for fundamental shifts. This is with regards to:

- The "Overview Effect"[14] and the "Chaos theory"[15] that envision a holistic understanding of world weather degeneration, a need for green pollution-free technology, and an organic-symbiotic embedment of our own technological progresses into the overall order of the universe
- Approaches to non-proliferation of arms and unnecessary wealth accumulation aimed at resources depletion only
- Recognition of parallel economies in Asia and Africa by the West

These fundamental changes are evident in the works of great personalities. Their work holds a fundamental shift in our approaches, in our attitudes, and in our belief systems. But that perhaps is not a new thing. This was always there in the ancient world, in India, particularly. The pointers forwarded by M. K. Gandhi, E. F. Schumacher, Amartya Sen, and Al Gore are a few amongst many.

The shift is only possible through a new temper best integrating the inner and outer pursuits of humanity – found in the complementarities of religion and science and in and through a balancing of two forces, one of the West and the other, of the East, to be seen in a temper of increasing harmony and assimilation and not in fight and dissension. As Albert Einstein famously put it, "Science without religion is lame, religion without science is blind." And as Werner Heisenberg concluded, "It is probably true quite generally that in the history of human thinking the most fruitful developments frequently take place at those points where two different lines of thought meet."

Notes

1 I have pursued this idea in *Concept of Complete Religion* (2006) at the generic level and in *Principles of Indian Architecture* (2008) in a particular historic context.
2 The discussion is based on *The Collected Works of Carl G. Jung.*
3 Based on the work entitled "Towards a psychology of Being" by Abraham Maslow.
4 Based on Ken Wilber's "The Spectrum of Consciousness."
5 Historian of Science, University of Berkley, California.
6 This is a wealthy suburb on Lake Michigan, USA, about twenty miles north of Chicago.
7 An alternating electric current (A.C.) mode has replaced the direct current now. It is popularly known as the A.C., which varies in strength, and it periodically reverses its direction. The frequency of an A.C. is the number of times the current passes through its zero value in the same direction in unit time.
8 This is a letter to Mr. Sturdy written from 228 W. 39th Street, New York City. The letter is re-mentioned in *The Master as I Saw Him* by Sister Nivedita, in Chapter 14 entitled "On death," page 303–305.
9 Please note that Einstein uses two different words: unity with regard to anthropic principle and a reality independent of that principle.
10 By "eternal" man Tagore possibly means the Cosmic Anthropic Principle, signifying total creative time.
11 Here Tagore mentions "standard" or a module. Swami Vivekananda in his works says, "the Macrocosm and the Microcosm are based on the same plan" – from essays on Cosmos: delivered in New York, Jan 26, 1896. Tagore means to say that there are through connections between the Eternal Man (the macrocosmic involutionary anthropic principle) and that of ours (the representative of microcosmic evolutionary anthropic principle). On the microcosmic side there are many, so there are many worldviews. This matches with the statement of Stephen Hawking.
12 It is a property of light giving it a kind of orientation vertically and horizontally in the lines of motion.
13 The idea comes from Schrödinger's parable of two kittens – Dead or alive (1995) – readers may refer John Gribbin in "Down to Earth," August 15, pp. 29–36.
14 It is a holistic idea of global warming and consequential degeneration of world weather explaining a holistic approach to earth's thermodynamic balance covering nature, ecology and human settlement patterns.
15 Idea developed by Nobel Laureate Ilya Prigogine in Chemistry of Nature.

Works cited

Aristotle. *Politics.* Ed. Peter L. Phillips Simpson. Chapel Hill: UNC Publishing, 1997.

Capra, Fritjof. *The Tao of Physics.* Colorado: Shambala Publications, Inc., 1975.

———. *The Turning Point- Science, Society and the Rising Culture.* London: Flamingo Press, 1990.

Chowdhuri, Asim. *Swami Vivekananda in Chicago: New Findings.* Kolkata: Adwaita Ashram.

Eyres, Harry. "Plato, Aristotle and the Commons." *Resurgence & Ecologist.* Issue 235. The Resurgence Trust (March–April 2006). Web.

Gribbin John, "In the Beginning: The Birth of the Living Universe". Ney York, Little Brown and co., 1995.

Livingston, Robert. *Sensory Processing, Perception, and Behavior.* New York: Raven Press, 1978.

———. *The Self-Organizing Universe.* New York: Pergammon, 1980.

Merchant, Carolyn. *The Death of Nature.* New York: Harper and Row, 1980.

Plato. *Republic.* Ed. G.M.A. Grube. Indianapolis, IN: Hackett Publishing, 1974.

Sen, Joy. *Concept of Complete Religion.* Kolkata: Cygnus, 2005.

———. *Principles of Indian Architecture.* Kolkata: Cygnus, 2007.

Tagore, R. *The Macrocosm and the Microcosm Are Based on the Same Plan.* Essays on Cosmos Delivered in New York. 26 January 1896.

13

RECONCILING FREE WILL AND DETERMINISM

An Indian appropriation of Benjamin Libet's neuroscientific findings[1]

Augustine Pamplany

Consciousness is such a fascinating mystery that both scientists and philosophers are equally interested in it. The neuroscientific interest in consciousness is a rather recent development. The neuroscientific presupposition is that consciousness is basically a neural phenomenon. Many others combine self-awareness and awareness with consciousness, denying the presence of consciousness in most animals and plants. Some philosophical traditions suggest that consciousness pervades the entire universe and can be found in animate and inanimate objects, only varying in the degree of its intensity. Thus the difficulties in grasping consciousness are obvious.

As human beings, the subjective inner life is unique to everyone, and it does differentiate he/she from the rest of the living beings. The volitional brain, free will, sense of a purpose in life, a knowing self, a reflecting mind, a soul that tend towards the Higher, etc. are aspects of one's unique subjective inner life. We know and understand only very little about the functions of our conscious will to act. Now, we do know that the physical brain also has got something to do with the manifestations of our conscious, subjective experiences. This chapter focuses on, perhaps, the most revolutionary experimental research to date into the nature of mind and free will as completed by the neuroscientist Benjamin Libet and his team. These findings have many things to say as to whether we are completely defined by the deterministic laws of nature or have some independence in making choices and actions. The neuroscientific claims of Libet have far reaching implications for ethics in general and bioethics in particular as it radically challenges the way we think about ourselves. Benjamin Libet, one of the leading neurophysiologists of our time, has taken a decisive step, a new approach to the so called mind-body problem. Based on experiments, he claimed that neurally based unconscious urges do precede conscious decisions. However, there is real evidence for true free will without unconscious neural causes. Such a position goes contrary to the conventional models of the operation of our free will. If our actions are unconsciously initiated, our moral responsibility and freedom of choice and free will are in danger. Understandably, Libet's findings raised a lot of hue and cry in the scientific and academic community. Although much has been said and written on Libet, I believe that the real ontological and metaphysical issues dormant in Libet still remain unaddressed. The questions thrown up by Libet, scientific as they are, are equally philosophical. There is much more at stake in Libet's findings than what is spelt out in the popular debate on Libet. An Indian philosophical look at Libet will be a pioneering direction in this field.

In the first part of this chapter, we deal with some philosophers and philosophical systems that have contributed to the debate on mind and matter. With this background, we shall present an objective presentation of the claims of Libet. Gradually, we orient our attempt to a more hermeneutical understanding of Benjamin Libet from an Indian philosophical approach to consciousness.

Historical overview of the mind-body problem

The question of free will has been a central issue since the beginning of philosophical thought. Unlike the ancient philosophers like Pythagoreans, Socrates, and Plato who attempted to reconcile an element of human freedom with material determinism[2] and causal law, Aristotle firstly argued convincingly for some indeterminism.[3] Though Aristotle stood for some indeterminism, he spoke about a causal chain back to a prime mover or first cause, and he elaborated the four possible causes – material, efficient, formal, and final. In *Metaphysics*, Aristotle makes the case for chance and uncaused causes, and in *Ethics* he shows our actions can be voluntary and so that we can be morally responsible. It is voluntary because, though they were predictable based on habit and character, character itself was developed freely in the past and was changeable in the future. Aristotle believed that whatever happens in our lives is necessary and should be regarded as coincidences. He also regarded it as the intersection of independent deterministic causal chains (Menn 275). Aristotle believed that many decisions that we make are quite predictable based on our habit and character. In his *Ethics*, Aristotle holds that that genuine moral virtue is not simply habituation to desire the right amount but involves choice, which involves deliberation or means-end reasoning, and moral virtue is not possible without the intellectual virtue of prudence or practical wisdom (Singh 127).

Although we can consider Aristotle as a non-determinist, nowhere does he discuss whether our decisions and actions are determined by circumstances beyond our control. However, he did claim that actions are compulsory when the cause is in the external circumstances and the agent contributes nothing. It is difficult to say whether Aristotle's claim that the cause of action is external implies that the agent contributes nothing and is therefore not free in the sense relevant to responsibility (ibid.).

Aristotle objected to the view that our actions are determined by our character in that it would deny moral responsibility. Though some aspects of our character might be innate and thus limit our responsibility, we are at least partially free to form our character. Even when our character determines our choices, we are now indirectly responsible for all those choices. Aristotle clearly believed that our deliberations involved chances between alternative possibilities. It is fully unto us whether to act or not to act, and this implies both the possibility to do otherwise and, conversely, moral responsibility for our actions.

Historically, a full-fledged discussion on the problem of free will was first held by the founding fathers of the Christian church. They identified a new problem in it as free will being in tension both with God's omniscience and with God's omnipotence. If God is omniscient, God must have foreknowledge of our choices, and if God is omnipotent, our actions must have been determined along with the rest of God's creation. St. Augustine came up with his notion of free will without underestimating the role of God in it. Augustine's classical reconciliation is that "humans are at least free to accept divine grace and redemption" (Markus 395). Augustine argued for free will but only as compatible with God's foreknowledge. That means God knows everything that happens in the world as well as that which is going to happen. It doesn't mean that our free will is restricted or limited. He says: "God must have given free will to man. God's foreknowledge is not opposed to our free choice" (Augustine 7).

Immanuel Kant's *Practical Reason* establishes freedom in a noumenal realm whose grounding principle is morality. Kant sees freedom as the condition for the moral law. He asks how we can think of the will as free and at the same time regard ourselves as subject to the moral law, i.e., as under obligation. Addressing this question, Kant invokes the idea of the two worlds, the sensible and the intelligible, to which he made appeal in *Critique of Pure Reason*.

> Insofar as I exercise the faculty of reason I have to regard myself as belonging to the intelligible world; insofar as I exercise my 'lower' faculties I am part of the world of nature, which is known through the senses.
>
> (Kant 176)

As a purely rational being, possessed of what Kant sometimes calls a "holy will,"[4] one's actions would be in perfect conformity with the principle of autonomy,[5] and the notions of obligation and the moral law would have no meaning here. They would similarly have no meaning if one is a purely sensuous being, for then everything one does would occur according to natural necessity, and there would be no sense in thinking that things ought to be otherwise. How is freedom possible? Kant explains,

All men think of themselves as having a free will. . . . Moreover, for purpose of action the footpath of freedom is the only one on which we can make use of Reason in our conduct. Hence to argue freedom away is as impossible for the most abstruse philosophy as it is for the most ordinary human reason.

(Kant 182)

Kant sees no difficulty in accepting the postulate of freedom because there is no contradiction in thinking of the will as free. Kant claims that freedom is the source of all values:

Freedom is, on the one hand, that faculty which gives unlimited usefulness to all the other faculties. It is the highest order of life, which serves as the foundation of all perfections and is their necessary condition. All animals have the faculty of using their powers according to will. But this will is not free. It is necessitated through the incitement of stimuli, and the actions of animals involve a *bruta necessitas*. If the will of all beings were so bound to sensuous impulse, the world would possess no value. The inherent value of the world, the *summum bonum*, is freedom in accordance with a will that is not necessitated to action. Freedom is thus the inner value of the world.

(Guyer 107)

Thomas Hobbes attempted to break completely with the past. He denied that there is any power in a human being to which the term will refers; what is commonly called will is but the desire in deliberating. Furthermore, he argued, only a human is properly called free, not his/her desires, will, or inclinations. To speak of liberty is not to make any suggestions about the determinants or absence of determinants of human's deliberations or decisions; it is to suggest that a human being is not externally constrained in his/her actions. There is, therefore, no contradiction in saying that human beings act freely and that his/her actions are also determined. Since all actions have causes and thus are necessitated, it is pointless to use "free" in the sense of "free from necessitation," as distinct from "free from compulsion" (Peters 416). However, we cannot accede to the minimalist understanding of will in Hobbes. He has just graded will as one among the desires or inclinations. Its value in guiding and regulating our desires and inclinations are not affirmed adequately by Hobbes.

For Hobbes, the idea that one could ever do otherwise was a contradiction and nonsense.

I hold that ordinary definition of a free agent, namely that a free agent is that which, when all things are present which are needful

153

to produce the effect, can nevertheless not produce it, implies a contradiction and is nonsense; being as much as to say the cause may
be sufficient, that is necessary, and yet the effect shall not follow.

(Hobbes 87)

But Hobbes may also be identified as the modern inventor of compatibilism,
the idea that necessary causes and voluntary actions are compatible.

When first a man has an appetite or will to something, to which
immediately before he had no appetite nor will, the cause of his will
is not the will itself, but something else not in his own disposing.
So that whereas it is out of controversy that of voluntary actions
the will is the necessary cause, and by this which is said the will is
also caused by other things whereof it disposes not, it follows that
voluntary actions have all of them necessary causes and therefore
are necessitated.

(Hobbes 98)

Determinism and compatibilism are a few concepts which we need to
clarify in this context. Determinism is the view that all current and future
events are causally necessitated by past events combined with the laws of
nature. Determinism is a broad term with a variety of meanings. Corresponding to each of these different meanings arises a different problem of
free will. Causal determinism is the thesis that future events are necessitated
by past and present events combined with the laws of nature as asserted
by Plato. The problem of free will, in this context, is the problem of how
choices can be free, given that what one does in the future is already determined as true or false in the present (ibid.). Theological determinism is the
thesis that there is a God who determines all that humans will do, either
by knowing their actions in advance, via some form of omniscience or by
decreeing their actions in advance. The problem of free will, in this context,
is the problem of how our actions can be free if there is a being who has
determined them for us ahead of time (ibid.).

Compatibilists argue that the assumption of free will and determinism are
compatible with each other; this is opposed to incompatibilism, which is
the view that there is no way to reconcile a belief in a deterministic universe
with a belief in a concept of free will beyond that of a perceived existence.[6]

Compatibilists maintain that determinism is compatible with free will.
A common strategy employed by "classical Compatibilists," such as Thomas
Hobbes, is to claim that "a person acts freely only when the person willed
the act and the person could have done otherwise, if the person had decided
to" (Hobbes 85). Hobbes sometimes attributes such compatibilist freedom
to the person and not to some abstract notion of will, asserting, for example, that "no liberty can be inferred to the will, desire, or inclination, but

154

the liberty of the man; which consist in this that he finds no stop, in doing what he has the will, desire, or inclination to do" (*Encyclopedia* 417). In articulating this crucial condition, David Hume writes, "This hypothetical liberty is universally allowed to belong to every one who is not a prisoner and in chains" (Hume 127). To illustrate their position, compatibilists point to clear-cut cases of someone's free will being denied through rape, murder, theft, or other forms of constraint. They argue that "determinism does not matter; what matters is that individuals' choices are the results of their own desires and preferences, and are not overridden by some external (or internal) forces" (Honderich 24). Being a compatibilist does not imply borrowing a particular conception of free will; rather, to deny that determinism is at odds with free will.

The scientific awareness of the question of free will is noticeable in history. Scientifically it meant that the actions of the body, including the brain and the mind, break the laws of nature. Early scientific thought often portrayed the universe as deterministic, and some thinkers claimed that the simple process of gathering sufficient information would allow them to predict future events with perfect accuracy. Modern science, on the other hand, is a mixture of deterministic laws and indeterministic principles. Quantum mechanics predicts events in terms of probabilities and uncertainties. Current physical theories cannot resolve the question of whether determinism is true of the world. Even if indeterministic interpretation of quantum mechanics is correct, it may be applicable only to the microscopic phenomena. Some others object to this view, showing that many macroscopic phenomena are based on quantum effects, for instance, some hardware random number generators work by amplifying quantum effects into practically usable signals (Chopra 92). Quantum randomness and probability may imply the absence of traditional free will, since random actions cannot be controllable by a physical being claiming to possess such free will (Chopra 93). Granted this argument, only compatibilism would account for traditional free will.

Like physicists, biologists have frequently addressed questions related to free will. One of the most heated debates in biology is that of "nature versus nurture." It concerns the relative importance of genetics and biology as compared to culture and environment in human behavior. The view of most researchers is that many human behaviors can be explained in terms of humans' brains, genes, and evolutionary histories. People like Francis Crick assert with some assurance that we are nothing more than a bundle of neurons and that it will without doubt be possible to explain consciousness and soul neurologically. He writes in his *Astonishing Hypothesis*, "you, your joys and your sorrows, your memories and your ambitions, your sense of personal identity and freewill, are in fact no more than the behaviour of a vast assembly of nerve cells and their associated molecules . . . you're nothing but a pack of neurons" (Crick 3). On the other hand, there are philosophers as well as scientists who react to this kind of reductive materialism by inserting mind/soul with the so-called "explanatory gap" argument. However, it could be seen and is proven quite evidently that the core aspects of human experience are intimately related to the neurobiological process, as seen from numerous lesion studies and with the most modern techniques. This point of view raises the fear that such attribution makes it impossible to hold others responsible for their actions. Responsibility doesn't require behaviour to be uncaused, as long as behaviour responds to praise and blame. Moreover, it is not certain that environmental determination is any less threatening to free will than genetic determination (Chattopadhyaya 24).

Traditionally, the puzzles surrounding consciousness have been known as the "mind-body" problem. However, it is now almost clear that the aspect of body most closely involved with consciousness is the brain. There is extensive evidence suggesting that brain states have causal influences on conscious experiences, and experiences can have causal influences on the body and brain. However, the neural, the material, and the "stuff" of conscious experience seem to be very different (Velmans 3). In any case the evolutionary view of consciousness does not explain how the unconscious material world can give rise to the conscious mind. If the universe is inherently material in nature, it cannot allow for the conscious mind to emerge. Besides, if the universe is materially closed, it cannot make room for consciousness as an independent phenomenon. Hence, according to some scholars, it could be argued that the universe is not simply material alone, but spiritual also (Pradhan 101).

Benjamin Libet's neuroscientific findings

In this philosophical debate between materialistic determinism and free will, the neuroscientific findings made by Benjamin Libet and his team are envisaged to be offering a moderate alternative. The major focus of the research by Libet and his team is the temporal relation between neural events and subjective experience. The startling finding here is that we unconsciously decide to act before we consciously decide to do so (Kosslyn X). Hence, voluntary acts seem to be not really free and spontaneous; rather, they seem to be preceded and triggered by unconscious events. There are electrical changes in the brain which really initiate the action before we consciously decide to act. It would mean that our "free" acts are rather consequences of unconscious events than acts caused by free agents. However, free will might still be relevant by having the capacity of vetoing the unconsciously initiated act (Libet 47–57).

In 1965, Kornbuber and Deecke had shown that self-paced voluntary acts were preceded by a slow electrical change recordable on the scalp at the vertex, which was termed the Readiness Potential (RP) (Kornbuber 1–17). In this experiment the actions of the subjects took place at regular intervals of 30 seconds. However, in the experiment by Libet and his team, this constraint on freedom of action was removed. The experiment presupposes the operational definition of free will in accordance with the common view that firstly there are no external controls, and secondly, the action is fully wanted, initiated and controlled by the subject.

In their experiment the subjects were asked to perform a simple flick or flexion of the wrist at any time they felt the urge or wish to do so. The emergence of the RP in these acts was at 550 msec. before the activation of the muscle. Now, it was important to understand exactly at which point of time the conscious wish to act would emerge. As Libet puts it:

> My question then became: 'when does the conscious wish or intention (to perform the act) appear?' In the traditional view of conscious will and free will, one would expect conscious will to appear before, or at the onset, of the RP, and thus command the brain to perform the intended act. But an appearance of conscious will at 550 msec. or more before the act seemed intuitively unlikely. It was clearly important to establish the time of the conscious will relative to the onset of the brain process (RP); if conscious will were to *follow* the onset of RP that would have a fundamental impact on how we could view free will.
>
> (Libet 49)

To measure the temporal relation between the RP, the conscious wish, and the act, Libet et al. developed a model of a much faster clock (see Figure 13.1). With this device they were able to get the reported clock-time associated with the first awareness of the wish to move from the subjects (W). The reported conscious awareness of the wish to act was found to be taking place 200 msec. before the act.

They performed another test to check the reliability of these reports. In this, they delivered a weak electrical stimulus to the skin of the same hand without the knowledge of the subject but known to the experimental observers. The subjects accomplished the clock time for the stimulus with an error of only –50 msec. Libet has presented the results of many such groups of trials in the following diagram (see Figure 13.2):

The results show that the brain process to prepare for this voluntary act began about 400 msec. before the appearance of the conscious will (W) to act. Libet suggests that the actual difference in times is probably greater than the 400 msec. because the actual initiating process in the brain probably starts somewhere in an unknown area which activates the supplementary motor area in the cerebral cortex. The source of RP is the supplementary motor area. The conclusion is that our conscious voluntary acts are actually initiated unconsciously.

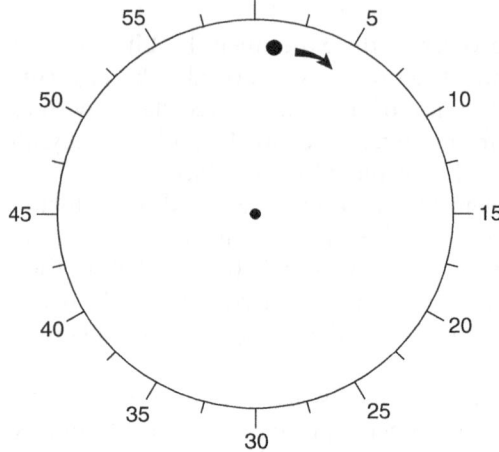

Figure 13.1 Oscilloscope "clock." Spot of light revolves around periphery of screen, once in 2.56 sec. (instead of 60 sec. for a sweep-second hand of a regular clock). Each marked off "second" (in the total of 60 markings) represents 43 msec. of actual time here. The subject holds his gaze to the centre of the screen. For each performed quick flexion of the wrist, at any freely chosen time, the subject was asked to note the position of the clock spot when he/she first became aware of the wish or intention to act. This associated clock time is reported by the subject later, after the trial is completed.

Source: B. Libet (1999), "Do we have free will?", *Journal of Consciousness Studies*, 6, pp. 47–57.

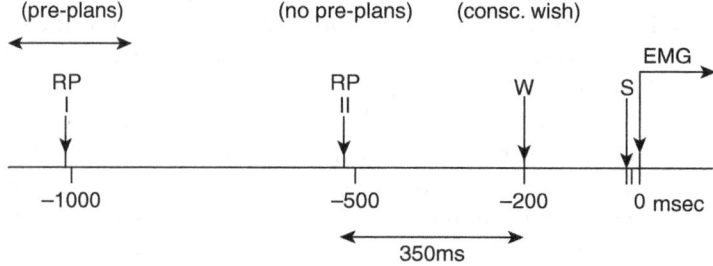

Figure 13.2 Diagram of sequence of events, cerebral and subjective, that precede a fully self-initiated voluntary act. Relative to 0 time, detected in the electromyogram (EMG) of the suddenly activated muscle, the readiness potential (RP) (an indicator of related cerebral neuronal activities) begins first, at about –1050 msec. when some pre-planning is reported (RP I) or about –550 msec. with spontaneous acts lacking immediate pre planning (RP II). Subjective awareness of the wish to move (W) appears at about –200 msec., some 350 msec. after onset even of RP II. However, W does appear well before the act (EMG). Subjective timings reported for awareness of the randomly delivered S (skin) stimulus average about –50 msec. relative to actual delivery time. ("Introduction" ix-xxiii, B. Libet, 1999, 'Do we have free will?', *Journal of Consciousness Studies*, 6, pp. 47–57.)

Is our conscious will merely a spectator in our voluntary acts? Is there any significant role for the conscious will? Libet does not rule out a role of the conscious will. Instead, he redefines the role of the conscious will:

> conscious will (W) does appear about 150 msec. before the muscle is activated, even though it follows onset of the RP. An interval of 150 msec. would allow enough time in which the conscious function might affect the final outcome of the volitional process. . . . Potentially available to the conscious function is the possibility of stopping or vetoing the final progress of the volitional process, so that no actual muscle action ensues. *Conscious-will could thus affect the outcome* of the volitional process even though the latter was initiated by unconscious cerebral processes. Conscious-will might block or veto the process, so that no act occurs.
>
> (Libet 51–52)

At this juncture, Libet asks whether the conscious veto function might perhaps also be originated unconsciously. However, Libet is inclined to think that the veto may not be the result of unconscious processes. Other scientists have tried to give a different role for conscious will within the experimental frame of Libet. Neuroscientist Velmans suggested that unconscious processes are offering a genuine choice to the conscious will to veto or not, and this choice would therefore still be a matter of conscious will ("Is Human Information Processing Conscious?" 651–669). Libet does not agree with this interpretation. After all, in Velmans' interpretation of the events, free will does not meet the condition that it should exercise a control function: it is only becoming aware of a preceding unconsciously initiated choice. Free will has no control over these unconscious functions.

> The conscious veto is a *control* function, different from simply becoming aware of the wish to act. There is no logical imperative in any mind – brain theory, even identity theory, that requires specific neural activity to precede and determine the nature of a conscious control function. And, there is no experimental evidence against the possibility that the control process may appear without development by prior unconscious processes.
>
> (Libet 53)

The decision of vetoing does imply an awareness of the event, but the awareness of the veto is different from the content of the veto, that is the decision to exercise the veto. "The awareness of the decision to veto could be thought to require preceding unconscious processes, but the content of that awareness (the actual decision to veto) is a separate feature that need not have the same requirement"(ibid.)

Although free will cannot not be equated with consciousness, Libet studies the neural phenomenon of consciousness to understand the role of free will. Given the veto role of the conscious will, the door is still open for the traditional notions of free will. But our consciousness of having made a decision is to be considered as a result of the unconscious brain processes. Our free will is not initiating the conscious decision, but it is the result of preceding unconscious brain processes. Kosslyn assesses Libet rightly, as he says, "Libet has made a fundamental discovery. If the meaning of mental events is as he describes, then we not only have 'free will' in principle – but we also have the opportunity to exercise free will" (Kosslyn xiv).

It is to be noted again that, in Libet, the conscious intention emerges after the RP and before the neural command for action. Some of the opponents of Libet who have attempted re-interpretation of the results of Libet have no hesitation in accepting the position of Libet that voluntary acts are non-consciously initiated. Although the real meaning of the RP and its temporal relation to the voluntary act is still to be determined fully, the prior occurrence of the RP before the volition seems to imply naturally that the action is initiated by the brain before the involvement of the mind. For Libet, there cannot be a free will in the sense of an ultimately responsible cause that is initiating the act.

Indian appropriation of Libet's findings

Although compatibilism seems to be a simple conclusion of Libet's neurological findings, a deeper critical scrutiny crossing the contours of an experiment-specific methodology and drawing on the postmodern academic trend of interdisciplinarity may provide a hermeneutical domain from which to reconceive the philosophical problem of moral responsibility in a scientifically tangible manner. Libet's experiment is the first and the most significant step towards a meaningful scientific analysis of the neurophysiological basis of conscious volition. Although this pioneering experimental work may not yield a comprehensive new paradigm altogether, redacting the metaphysical prejudices of the philosophical interpretations of this data could set the trend towards new directions and new assumptions of research in this field. In this section we shall subject the philosophical assumptions of Libet's findings to a critical scrutiny and suggest an alternative ontological grounding for the interpretation of his data.

What is often overlooked in a compatibilist presentation of Libet's findings is that, in his data, there is a break in the neurally deterministic causal chain of connections. Hence Libet's research is out of step with compatibilism, as argued by several authors.

Klein observes that, as the conscious veto in Libet is not preceded by unconscious neural deterministic precursors, his view of free will is antithetical to determinism. In fact, Libet's view requires a break with natural law. This break with natural law is warranted in classical mechanics ontology, with dualism in the philosophical background. Although Libet is keen to avoid *ad hoc* assumptions and philosophical hypotheses in his research and rebuts his critics on several occasions on the use of such premises, he does not avert a paradoxical fall into an overarching dualistic ontology in the interpretation of his results. On the scientific side, this dualism shows itself in the "out of nothing" appearance of the conscious veto or trigger by delinking it with the neural unconscious process. On the philosophical side, this has led Libet to an old-fashioned Cartesian dualism. This is explicitly articulated in Libet's imaginary dialogue with René Descartes. But this dualistic direction does seem to represent neither scientific nor philosophical progress from his experimental findings. The isolation of the conscious veto is one of the weakest points in Libet's theory. As Clark has rightly observed,

> Libet's case . . . for the independence of the conscious veto from neural activity is not that there is positive evidence for it, but merely that logical and empirical consideration don't rule it out. But even this is too strong a claim.
>
> (Acton 576)

Given Libet's findings, should we not be looking more radically at the compatibilist paradigm of reconciling determinism and free will? Although the compatibilist account of free will adequately fits our sense of morality etc., such an appropriation would represent a minimalist philosophical approach, distinctly preferring the simple and the empirical over the depth and complexity of the point under debate.

The popular libertarian-compatibilist debate on moral responsibility has the classical ontology of the mechanical philosophy of nature in its background. However, a new ontology informed by the interdisciplinary data and intercultural worldviews might settle the traditional debate on moral responsibility in different ways. Furthermore, Libet's findings might provide a pioneering neurological foundation for such a conceptual change. It is our contention that artificial reconciliation between the unconscious Readiness Potential and the conscious veto in Libet could be resolved if the cosmic understanding of consciousness and the mind-matter unity of the Indian tradition are alluded to.

Although many Indian neuroscientists have responded to Libet, there is no adequate attempt made so far to appropriate the findings of Libet within the framework of the traditional Indian perspectives on consciousness, mind, etc. The experimental findings of Western neuroscience are to some extent complemented by the intuitive insights of India on consciousness. My contention in this chapter is that the findings of Libet point to the macroscopic or cosmic foundations of consciousness, and a cosmic perspective on Libet is analogically in consonance with the Indian perspective on consciousness which very much emphasizes the cosmic outreach of consciousness. In India, consciousness is not an exclusive priority of the human person. Indian understanding of the very concept of existence reaches out downward to the insentient elements and upward to the spiritual and divine principle.

The ancient cosmogonic theories, like the metaphor of *hiranyagarbha* (golden egg), presented an organic image of the universe whereby every bit of cosmos is related to every other part of the universe. However, the element of consciousness was very much missing in the more ancient explanations concerning the origin and nature of the cosmos. The Vedic cosmogony thus stood in need of the de-materialization of the first cause of the universe. And this is done with the introduction of the profound metaphor of *Purusha*. *Rigveda* 10.90 speaks of the origin of the universe from the sacrifice of primordial being called *Purusha*. *Purusha* to some extent is a conscious and personal principle. By attributing the very origin of the entire cosmos to a primordial conscious principle, the Indian mindset has ruled out the mind-matter dualism which had been characteristic of the Western philosophy.

Sacrifice being the focus of the whole Vedic literature, Vedic thinkers naturally conceived the origin of the cosmos as emerging from the sacrifice of a giant primordial being called *Purusha*. In the sacrifice, the organs of the *Purusha* were cut out by gods and those parts became the different parts of the world.

10.90.1. A thousand heads had the *purusha*

> A thousand eyes, a thousand feet;
> Encompassing the earth on every side,
> He exceeded it by ten fingers (breath).

10.90.2. *Purusha* is the whole universe,

> What was and what is yet to be;
> He is the Lord of Immortality
> And of what grows was by (eating) food.

10.90.13. From his mind the moon was born,

> And from his eye the sun,
> From his mouth Indra and the faire,
> From his breath the wind was born.

India has looked at mental activity as a many-layered entity. The basic level is consciousness, known as *prajna* or *chetana*. On this base, the sensory and motor systems (the *jnanendriyas* and the *karmendriyas*) create the mind (*manas*, *chittam*), which responds to the impulses from the senses (*indriyas*). The response could be positive or negative, leading to attraction (*raga*) or repulsion (*dwesa*). These build up emotions and create enjoyment (*sukham*) and distress (*dukkham*), resulting in desire (*kama*), anger (*krōdha*), avarice (*lobha*), and hatred (*dwesa*). All these are still at the mind level (*manas*) and the whole system is included in the *indriyas* (Ramamurthi 324).

Above this mind or *manas* or *chitta* operates intelligence (*buddhi*) which is capable of discrimination and also capable of imposing its decisions on the mind. Above this level is the feeling of "I"ness or self-awareness (*aham-kara*). This "I"ness leads to the feeling of possession, "mine" (*mamatva*), the sense of ownership of the whole body and of all the mental activities including all emotions.

It is a pity that for two centuries the Europeans did not care to study the Indian concepts of the mind and the body even though they were ruling us. If that had been done, René Descartes would not have created the dual mind-body problem nor would have made the now popularized statement *Cogito, ergo sum* (I think: therefore I am). He would have said "I am: therefore I think," and neuroscience or European philosophy perhaps would not have lost almost a century in their investigation of the mind. However, these days, many studies are being undertaken by Europeans on Indian concepts of mind and body.

The rich allusiveness of the Indian vision is to some extent absorbed in the modern insights in physics and chemistry, which further supports our juxtaposing of Libet and the Indian vision on a cosmic grounding. In 1925 Schrödinger and Heisenberg studied the behavior of matter more deeply, and the result of their work was virtually to abolish the pure materialistic concept of matter. We were used to thinking of matter as the basic thing and of energy as its property, but we now think of energy as the basic thing. The view at present is that perhaps electrons and protons have certain amount of free will, so that their behavior, even in theory, is not wholly subject to law (McCrone 183). The principle of uncertainty cuts at the root of the traditional physics in which position and velocity are fundamental. This is the breaking down of physical determinism and favouring the rehabilitation of free will.

Haeckel, the modern natural scientist, said:

> Without the assumption of an atomic soul the commonest and the most general phenomena of chemistry are inexplicable. Pleasure and pain, desire and aversion, attraction and repulsion must be common to all atoms of an aggregate. For the movements of atoms, which must take place in the formation and dissolution of a chemical compound can be explained only by attributing to them sensation and will.
>
> (Handy 202)

Explaining his statement further in his book *The Riddle of the Universe*, he says:

> I explicitly stated that I conceived the elementary psychical qualities of sensation and will which may be attributed to atoms to be

163

unconscious – just as unconscious as the elementary memory which I, in common with the distinguished psychologist E. Wald Hering, consider to be a common function of all organized matter or more correctly the living substance.

(Acton 575)

Bishop Berkeley maintained that

we have no right to assume that matter is different from mind. Matter outside mind produces ideas inside our minds. Causes must be of like nature to their effects. Thus Berkeley argued that matter must be of the same nature as an idea.

(Acton 576)

To say that mind cannot influence matter now becomes as absurd as to say that mind cannot influence ideas. The integral Advaitism of Aurobindo conceives of no fundamental opposition between spirit and matter, the Brahman and Cosmos, for matter is spirit and cosmos is divine – matter is in fact the lower form of the spirit.

It must be noted here that the latest findings of new physics also uphold the same truth that there is a fundamental unity in the world. The quantum theory has abolished the notion of fundamentally separated objects and emphasizes an essential interconnectedness of the universe.

As we study the various models of subatomic physics, we shall see that they express again and again in different ways, the same insight that the constituents of matter and the basic phenomena involving them are all interconnected, interrelated and interdependent; that they cannot be understood as isolated entities but only as interrelated parts of the whole.

(Smoot et al. 125)

The six orthodox (*astika*) schools of thought in Hindu philosophy do not agree with each other entirely on the question of free will. For the Samkhya, for instance, matter is without any freedom, and soul lacks any ability to control the unfolding of matter. The only real freedom (*kaivalya*) consists in realizing the ultimate separateness of matter and self (Kapoor 344). For the Yoga school, only Ishvara is truly free, and its freedom is also distinct from all feelings, thoughts, actions, or wills and is thus not at all a freedom of will (Kapoor 356). The metaphysics of the Nyāya and Vasheshika schools strongly suggest a belief in determinism but do not seem to make explicit claims about determinism or free will (Kapoor 175).

Swami Vivekananda sums up the controversy over free will in the Hindu tradition:

> Therefore we see at once that there cannot be any such thing as free-will; the very words are a contradiction, because will is what we know, and everything that we know is within our universe, and everything within our universe is molded by conditions of time, space and causality. . . . To acquire freedom we have to get beyond the limitations of this universe; it cannot be found here.
>
> (Vivekananda 118)

Mimamsa, Vedanta, and the more theistic versions of Hinduism such as Shaivism and Vaishnavism have often emphasized the importance of free will. The doctrine of *karma* in Hinduism requires both that we pay for our actions in the past and that our actions in the present be free enough to allow us to deserve the future reward or punishment that we will receive for our present actions (Kapoor 356).

Buddhism accepts both freedom and determinism but rejects the idea of an agent and thus the idea that freedom is a free will belonging to an agent. According to Buddha, "There is free action, there is retribution, but I see no agent that passes out from one set of momentary elements into another one, except the 'connection' of those elements" (Bettany 274). Buddhists believe in neither absolute free will nor in determinism. Buddhism preaches a middle doctrine, named *pratitya-samutpada* in Sanskrit, which is often translated as "inter-dependent arising." It is part of the theory of karma in Buddhism (Bettany 277). The concept of karma in Buddhism is different from the notion of karma in Hinduism. In Buddhism, the idea of karma is much less deterministic. The Buddhist notion of karma is primarily focused on the cause and effect of moral actions in this life, while in Hinduism the concept of karma is more often connected with determining one's destiny in future lives.

In Buddhism it is taught that the idea of absolute freedom of choice (i.e., that any human being could be completely free to make any choice) is foolish, because it denies the reality of one's physical needs and circumstances. Equally incorrect is the idea that we have no choice in life or that our lives are pre-determined. To deny freedom would be to undermine the efforts of Buddhists to make moral progress (through our capacity to freely choose compassionate action) (*Sacred Books of the East* 51). *Pubbekatahetuvada*, the belief that all happiness and suffering arise from previous actions, is considered a wrong view according to Buddhist doctrines. Because Buddhists also reject agenthood, the traditional compatibilist strategies are closed to them as well. Instead, the Buddhist philosophical strategy is to examine the metaphysics of causality. Ancient India had many heated arguments about the nature of causality, with Jains, Nyayists, Samkhyists, Cārvākas, and Buddhists all taking slightly different lines. In many ways, the Buddhist position is closer to a theory of "conditionality" than a theory of "causality," especially as it is expounded by Nāgarjuna in the *Mūlamadhyamakakārikā* (Bettany 286).

Libet talks about unconscious or preconscious initiation of the volitional process, especially the "free" choices that we take in our day to day life. On a speculative ground we could say that, in a way, Libet's findings imply the idea that matter is conscious. Coming to Indian philosophy, we can find ideas similar to this.

The *Yoga-Vasista* is a rich and complex philosophical "poem" written in Sanskrit by an unknown author. It is notable for its eloquent praise of self-effort and inquiry or analysis, and for its severe disparagement of the notion of fate. It views consciousness as characterizing all living forms (including plants and insect life), being atomic, and analogous to the emergence of waves and whirlpools in water; it therefore grapples with what today would be called the problems of reductionism and emergentism. There is a reference to atomism also here in that life is seen as made up of atoms – but possibly endowed with memory or some special ingredient. It also says that consciousness emerges in some similar way from those atoms without being forced by something beyond them (Narasimha 246). Here also we can see a form of conscious matter.

Conclusion

After having outlined the philosophical setting of the debate between free will and determinism, we discussed the neuroscientific experimental findings of Benjamin Libet. We observed that the tension between unconscious origination of the Readiness Potential and the conscious origin of Veto in Libet could be resolved if we are to take recourse to a more cosmic understanding of consciousness. Indian perspectives on consciousness provide us with sound conceptual grounding for such a vision. Our attempt has been just introductory, and it has been only suggestive of some directions for in-depth research in this line, which will be quite rewarding in settling some of the perennial issues in the philosophy of mind.

Notes

1 This chapter is a revised version of an earlier draft published in Augustine Pamplany (ed.), *Mastery Meets Mystery: Intersecting Science, Philosophy, Religion and Culture* (New Delhi: Serials Publications, 2015), pp. 247–274.
2 Determinism is the view that every event, including human cognition, behavior, decision, and action, is causally determined by an unbroken chain of prior occurrences.
3 Indeterminism is a philosophical position that maintains that some form of determinism is incorrect: that there are events which do not correspond with determinism.

4 In the ethics of Kant, a holy will would be spontaneously inclined to obey the dictates of morality; it would not have to struggle against countervailing desires and inclinations or the innate human tendency to evil. This moral perfection is not a possible achievement for human beings.

5 An autonomous will is entirely self-legislating: the moral obligations by which it is perfectly bound are those which it has imposed upon itself while simultaneously regarding them as binding upon everyone else by virtue of their common possession of the same rational faculties.

6 Again some others accept the existence of free will along with an assumption of indeterminism to some extent. They are known as libertarians.

Works cited

Acton, H.B. "Berkeley, George." *Encyclopedia of Philosophy*. Ed. Donald M. Rorchert. New York: Macmillan Reference, 1996. 575–576.

Augustine, St. *On Free Choice of the Will*, Book 2. 7. Web.

Bettany, G.T. *The Buddhist Doctrine: Encyclopedia of World Religion*. New Delhi: Victory Books International, 1991.

Chattopadhyaya, D.P. "Physics, Body-Mind and the Beyond." *Science and Metaphysics: A Discussion on Consciousness and Genetics*. Eds. Sangeetha Menon, Anindya Sinha, and B.V. Sreekantan. Bangalore: National Institute of Advanced Studies, 2002. 4.

Chopra, Deepak M.D. *Quantum Healing: Exploring the Frontiers of Mind/Body Medicine*. New York: Bantam Books, 1990. 92.

Crick, Francis. *Astonishing Hypothesis: The Scientific Search for the Soul*. New York: Simon & Schuster, 1995. 3.

"Free Will." *Stanford Encyclopaedia of Philosophy*. 29 October 2010. Web.

Guyer, Paul. *Kant on Freedom, Law and Happiness*. London: The Press Syndicate of the University of Cambridge, 2000. 107.

Handy, Rollo. "Haeckel, Ernst Heinrich." *Encyclopedia of Philosophy*. Ed. Donald M. Rorchert. 2nd ed. New York: Macmillan Reference, 2006.

"Hobbes, Thomas." *Encyclopedia of Philosophy*, 417.

Hobbes, Thomas. *Leviathan*. Ed. John Plamenatz. London: Fontana, 1969.

Honderich, Ted. "Compatibilism and Freedom." *Encyclopedia of Philosophy*. Ed. Donald M. Rorchert. 2nd ed. New York: Macmillan Reference, 2005.

Hume, David. *A Treatise of Human Nature*, vol. 2, Book 2, Part 3. Great Britain: J.M. Dent & Sons Ltd., 1961. 127.

Kant, Immanuel. *Critique of Pure Reason*. Trans. Max Muller. New York: Anchor Book, 1966. 176.

Kapoor, Subodh. (ed.). *Encyclopedia of Indian Heritage*. New Delhi: Cosmo Publications, 2002.

Kornbuber, H., and L. Deecke. "Hirnpotentialanderungen bei Willkurbewegungen und passive Bewegungen des Menschen: Bereitschaftspotential und reafferente Potentiale." *Pfluegers Arch Gesamte Physiol Menschen Tiere*. 284: 1–17. 1965.

Kosslyn, S.M. Foreword to Benjamin Libet, *Mind Time – The Temporal Factor in Consciousness*. Cambridge: Harvard University Press, 2004. x.

Libet, Benjamin. "Do We Have Free Will?" *Journal of Consciousness Studies*. Vol.6 (1999).

Libet, Benjamin et al. "Editor's Introduction: The Volitional Brain – Towards Neuroscience of Free Will." *Journal of Consciousness Studies*. Vol.6 (1989): ix–xxiii.

Markus, R.A. "Augustine S.T." *Encyclopedia of Philosophy*. Ed. Donald M. Rorchert. New York: Macmillan Reference, 2006.

Mc Crone, John. *Going Inside: A Tour Round a Single Moment of Consciousness*. New York: International Publishing Corporation, 2001. 183.

Menn, Stephen. "Aristotle." *Encyclopedia of Philosophy*. Ed. Donald M. Rorchert. New York: Macmillan Reference, 2006.

Narasimha, Roddam. "A Metaphysics of Living Systems: Reductionism and Emergentism in Yoga-Vasistha." *Science and Metaphysics: A Discussion on Consciousness and Genetics*. Eds. Sangeetha Menon, Anindya Sinha, and B.V. Sreekantan. Bangalore: National Institute of Advanced Studies, 2002. 245.

Peters, R.S. "Hobbes, Thomas." *Encyclopedia of Philosophy*. Ed. Donald M. Rorchert. New York: Macmillan Reference, 2006.

Pradhan, R.C. "Why Consciousness Cannot Be Deconstructed: Towards a Positive Theory of Consciousness." *Science and Metaphysics: A Discussion on Consciousness*. Eds. Sangeetha Menon, Anindya Sinha, and B.V. Sreekanthan. Bangalore: National Institute of Advance Studies, 2002. 101.

Ramamurthi, B. *Sacred Books of the East*. Ed. Max Muller. Vol. X. New Delhi: Motilal Banarsidass Publishers Pvt Ltd., 1962. 51.

———. "Consciousness – Elusive or Can One Grasp It?" *Science and Metaphysics: A Discussion on Consciousness and Genetics*. Eds. Sangeetha Menon, Anindya Sinha, and B.V. Sreekantan. Bangalore: National Institute of Advanced Studies, 2002. 324.

Singh, Rita. *Aristotle's Philosophy of Science*. New Delhi: Global Vision Publishing House, 2003. 127.

Smoot, Robert C. et al. *Chemistry: A Modern Course*. Toledo: Merrill Publishing Company, 1990. 125.

Velmans, Max. "Is Human Information Processing Conscious?" *Behavioural and Brain Science*. Vol.3 (1991): 651–669.

———. *Understanding Consciousness*. London: Routledge, 2000. 3.

Vivekananda, Swami. *The Complete Works of Swami Vivekananda*. Kolkata: Advaita Ashrama. Web.

14

SCIENCE AND SPIRITUALITY FROM THE PERSPECTIVE OF THE *MAHABHARATA*

Sitansu Chakravarti

There is an ongoing dialogue on the relationship between science and spirituality. If we get a handle on the meaning of the important expression "spirituality," we will certainly be in a better shape to see how the interaction between these two areas of human involvement can be viewed as welcome in practice. I will first attempt to see the relation between science and spirituality on the strength of the intuitive sense we have of the expression via a comparative view of the two diverse disciplines of rigorous reasoning viz. science and philosophy.

Science, philosophy, and spirituality

Scientific reasoning is methodical; it does not take anything for granted to begin with. Thus, doubt is its starting point. The first philosopher in the West who started with "methodological doubt" is René Descartes, and he is hailed as the father of modern philosophy for having followed the spirit of science in philosophy. The achievements arising out of the pursuit of science are often credited with elevating the methodology of scientific thinking to the ideal model for all serious thinking. However, philosophical reasoning, rigorous as it is, does not pursue the scientific model. One reason is that, whereas physics deals with the ontological nature of things, this may not be the only or even the main concern of philosophy. The subject matter of physics is never values; I do not know if there is any room for unconditional *anandam* or *mangalam* in quantum mechanics. The philosophical systems pertaining to spirituality take some values for granted as matters of faith not amenable to scientific ontological proofs. *Mangalam* is one such concept which is not to be taken only in the consequentialistic setting for it to emerge as the consequence of actions undertaken, as happens with building hospitals or schools, but an intrinsic value which is taken as the essence of human beings and inspires them to such actions. It is a value that the spiritual person sees embedded even when life is otherwise full of challenges; it

169

is already there and is inspiration for undertaking actions toward welcome changes in life. The *Gayatri* Mantra may be looked upon as directing to this value which took root with the origin of the universe and toward which the principle responsible for the process of the origin of the universe leads us all. It is the changeless Shiva, or *mangala*, the ingredient of the creation process in a very strange, certainly poetic, way, while Shakti leads us on to the *Shiva-loka*, the state of *mangalam*, which is the telos of life.

Thus, saying that consciousness is pervasive, as one may at times attempt to interpret quantum mechanics as doing, needs to be supplemented by a lot of further discussion to cover the whole dimension of truth in so far as it incorporates values. Neither do we reach this end point of the telos in life with the aid of scientific reasoning nor Sri Aurobindo's vision of the next step in the process of evolution leading to spiritual fulfillment. When Mother of Sri Aurobindo Ashram asks her devotees "Are you ready?" she does not mean that we start taking courses in quantum mechanics. What she means, I think, is that we may mentally prepare ourselves for the next step in the evolutionary process that Sri Aurobindo had talked about. For that we have built the right kind of attitude in us, conducive to the change to take shape while we participate in welcoming as well as precipitating the process.[1]

This value is spiritual and is the foundation of the ethical values in life. Spiritual pursuit is associated with diverse philosophical systems. One does not have to be a master philosopher to attain the height of spirituality, nor is spirituality connected with a system necessarily cancelled when faults are shown to exist with the philosophical reasoning involved there. Cancellation of the ontological underpinnings of a spiritual school by philosophical or scientific reasoning, for example, does not reduce the spiritual stalwarts belonging there to non-entities. Even if the Advaita metaphysics is, for example, "proved" wrong from the Visistadvaita standpoint, Ramana Maharshi would stand out, although his attainments would fail to be amenable to the Advaita interpretation. The above also goes to show that spirituality is not as intimately connected, if at all, to ontology as science is. Philosophical reasoning is neither a necessary nor a sufficient condition for attainment of the spiritual state. For, even with "wrong" reasoning, one may have reached the goal, while one expert in the reasoning process may have missed out on it. However, lack of the needed skill in reasoning acts as an impediment to one's scientific status. To carry on the train of thought above, one may not have attained the spiritual state simply having worked out quantum mechanics either, which amounts to unveiling the true ontological status of the universe with use of the needed mathematical skill. It is not difficult to imagine a person excelling in quantum mechanics to be mean and selfish even when acclaimed as an outstanding physicist, thus failing to be spiritual. Having the right knowledge here certainly does not amount to attainment of the desired spiritual state. Neither is this knowledge gained by scientific pursuit a necessary condition for spirituality, as borne out by examples of

Ramana Maharshi and several other figures, although spirituality is a necessary condition for pursuit of science in conformity with human values, a demand not made by science per se.

Act and attitude *dharmas*

We have not yet made the meaning of "spirituality" clear enough by pursuing our intuitive sense of the expression while observing how spirituality fares in the context of philosophical as well as scientific reasoning. Perhaps the *Mahabharata* may be of help here. The *Mahabharata* distinguishes between two senses of dharma in ethics, distinct and yet intimately connected: the *act dharmas*, which refer to the individual acts of morality, and the *attitude dharmas* signifying the attitudes proper for performance of the acts demanded by situational needs.[2] The needs are addressed ultimately on attitudinal constraints such as sympathy which brings in its wake other related attitudes, like forgiveness, truth, and control of greed, as mentioned by the *Mahabharata*, not only for initiating an act in a situation but also for the way it is to be performed. All these virtues mentioned can be clustered around the virtue of harmony, leading one to the unfolding of one's real nature in harmony with oneself, the society, and the ecology we all are situated in.[3] The more one is established in this virtue, the more one is expected to act in the ethical way. Being absolutely established in this virtue of harmony I would say is the spiritual goal of life, taken in its secular sense, irrespective of the fact of one's belief in God or understanding of quantum mechanics. With the dimension of inner and outer harmony missing, the spiritual dimension may be said to have been skipped. A strong believer in God indulging in mass killings or a perfect scientist working to develop a tool for destruction of the world is not a spiritual person.

Harmony and secular spirituality

In the *Gita* two words are used: *samatva* and *samya*. The first is used just before Sri Krishna starts describing the state the *sthitaprajna*, i.e., the state a person with steady wisdom is in, which is a state of becoming and not just of knowledge gathered through the pursuit of science.[4] Along with the expression *nirdvandva* used a little earlier,[5] meaning "free from discord," the sense of harmony is highlighted. This state of becoming, by the way, is missing in the scientist qua scientist who, in spite of his erudite knowledge, would not be a substitute for the holy man unless he turns into a holy man himself through practice of *nididhyasana*, i.e., absorption, on top of *manana*, the other name of intellectual interpretation. The *nididhyasana* again is not complete so long as the virtue of harmony has not set in. For, so long as one has not attempted being *samadarshi*,[6] with a sense of equanimity toward all, the ego lies there intact, and the spiritual dimension has not been attempted

171

either. The description of the *sthitaprajna* gives us a glimpse of the state of harmony in a secular setting without any ostensive reference to God. Such a person is ever after the good of all, including non-humans.[7] The mention of *samya* occurs in relation to someone who is established in Brahman.[8]

Taking *samatva* or *samya*, which goes hand in hand with equanimity towards all, as the basis for ethical acts and accepting the expressions as standing for spirituality, in the sense of an inclusive harmony surpassing the ontological dimension, we may say that ethics is grounded in spirituality.[9] The nearest approximation to the position of equanimity is perhaps in Kant's principle of universalizability of an act in ethics. The concept of equanimity in the *Gita* certainly implies universalizability at the attitudinal level, with an existential bearing to the concept of becoming, though not universalizability of an act that Kant highlights as a basic ingredient of his deontological theory of ethics. Kant's concept of universality is confined to the realm of ethics, where ethics fails to link itself to the spiritual dimension, unlike in the case of the *Gita*. Thus, the *Gita* has the advantage of accommodating the pluralistic dimension for the moral act, which does not have to be universalizable as demanded by Kant, without compromising at the same time with moral relativism, having accepted the universal attitude/s in the theory. It has the added advantage of providing a basis for the act, both regarding theory and practice, for one to inculcate in life toward its fulfillment through performance of the right acts guided by the attitude/s. The spiritual dimension here does not necessarily make the position of the *Gita* less secular, in that the state of the spiritual height is conceived here in absolutely secular terms, leaving room for its attainment in secular ways, as evidenced in the thoughts of the Samkhya school. There does not have to be an added ontological underpinning here either.

Conclusion

Thus, the sustaining force for ethics is spirituality. The vast literature on logical positivism has a significant dearth of writings on ethics in so far as logical positivism deals with the realm of science. Although the ethical and aesthetic values find their limited inclusion eventually in the system, they are not contained there in the unified frame of the spiritual dimension of attitudinal virtues not to be accommodated in the science of physics.

Science, we saw, is no substitute for spirituality, which is not (only) a search for the ontological state of the being but a pursuit of a value through the process of becoming. God is not a being in the Hindu system but is the value of unconditional *ananda*, or joy, and is nothing but Existence, Consciousness, and Bliss, which are not marks of a being. The *Gita* says that all our actions are to be performed in the spiritual mode toward the attainment of the height of the spiritual state. Scientific pursuit must follow this route in line with the other pursuits in life without ever posing itself as a substitute for spirituality. However, the way science, quantum mechanics specifically, is especially relevant to spirituality is via its thrust on the unity of the universe, more so when one views things in the light of an all-pervasive consciousness. This thrust toward unity has a tendency to lead to the spiritual thrust of attaining harmony within and without, although the latter may not be reduced to the thrust toward unity to be found in science. Relating oneself to this thrust toward unity and being established in it in actual life is what spirituality consists of. This is involvement in the process of an art not within the purview of science. As Mundakopanishad says: *Nayam atma pravacanene labhyah, na medhaya, na bahuna shrutena.*[10] In other words, listening about Brahman or arguing about it is not necessarily a spiritual mode, but attempting to relate oneself to one's real nature either apart from these involvements, or along with them, is. Quantum mechanics is at most in the business of analyzing at the meta-level the relating process, but it is not itself involved in the process of relating itself. Involvement in this art of the relating process is more akin to involvement in the art of writing poetry, composing songs, or painting, contrasted with the applied science of engineering a very efficient computer on the basis of quantum theory.

I conclude here by indicating that the virtue harmony may be viewed in a setting parallel to the Chomskyan linguistic model of the deep level of language we all are born with translating itself into the surface languages. The universal moral principle may be viewed via the virtue situated at the deep level we all are born with, being *sahaja*, i.e., innate. The diverse practices are but the various ways it translates itself in societies, with an invitation to individuals to be firmly embedded in the virtue at the surface level of consciousness, in the existential mode, through practice of the variant ways in societies. Severing ourselves from the virtue and taking to parochial means keeps us away from the journey of life, which after all is a spiritual journey, whether one believes in God or not.

All in all, there is no antagonism between science and spirituality. The former may be conducive to it, but never a substitute for it. Both can come together only when the former is included within the latter, not vice versa. That would signal the stage where science would be for the benefit of humanity, keeping ecology as well as well-being of all creatures in perspective, where greed would not be the driving force for human action.

Notes

1 We may safely take the value we have referred to as an ingredient of the first synthesis "in the long history of Indian thought" that Sri Aurobindo talks about, the Vedic synthesis, in other words, "of the psychological being of man in its highest flight and widest rangings of divine knowledge, power, joy, life and glory . . . pursued behind the symbols of the material universe into those superior planes which are hidden from the physical sense and the material mentality."

2 The *Mahabharata*, 5.35.56–7; 3.2.75–9. There is detailed discussion on the area in the book, *Ethics in the Mahabharata: A Philosophical Inquiry for Today*, Munshiram, 2006, New Delhi.

3 Sympathy is not enough by itself, even when accompanied by care and empathy, as Michael Slote suggests in *The Ethics of Care and Empathy* (Oxford: Routledge, 2007), for it must be guided by truth, both of which, viz. sympathy and truth, are included in harmony in so far as truth provides a direction to sympathy.

4 The *Gita*, 2/48.

5 Ibid., 2/45.

6 Ibid., 6/29.

7 "*sarva-bhuta-hite-ratah*," ibid., 5/25, 12/4.

8 Ibid., 5/19.

9 This is the lowest common denominator for the meaning of the expression which signals, in a secular way, the end for all different faiths toward a common goal. However, if science does not live up to it, then to that extent it is away from spirituality.

10 *Mundakopanishad*, 3/2/3.

15

REVISITING CONCEPTS OF HEALTH AND DISEASE

Evolution, philosophy, and integration

Bhushan Patwardhan[1]

Concept of health

The concept of health often revolves around a few key words like well-being, wellness, and happiness. The etymology of the term *health* is very interesting. *Health* in Old English actually meant "wholeness, sound or well." It seems to have its origin in the proto-Germanic word *hailitho*, meaning "whole, uninjured, of good omen." An old Norse term, *heill*, means "healthy," and *hælan* means "to heal." Health also denotes prosperity, happiness, welfare, preservation, and safety. Often the terms health and wellness are used together in Western culture. In Sanskrit, the word for health is *aarogya*, which indicates an absence of disease. A more appropriate term for health as used in Ayurveda is *swasthya*.

In many cultures and traditions the proverb "health is wealth" is popular. For every individual, health is crucial for living well. Everyone hopes and aspires to be healthy, naturally. However, the concept of health is not limited to individuals but also extends to community health, public health, human health, animal health, plant health, and environmental health. The concept of health even extends to specific ecosystems like oceans, rivers, habitats, towns, cities, and nations. Health is equilibrium, a balance, and a state of harmony. When we lose it, we know that it existed. It is easier to know illness but very difficult to know wellness, and consequently health is described and discussed mostly in relation to disease, illness, or sickness.

Many times the term *health* is equated with the term *medicine*. This erroneous perception of health is based on disease models, which might have led to the current disease focus. Today, health is considered a commodity which can be acquired through medicines and health care. In reality, health is a positive and dynamic concept. It can be attained, acquired, and maintained only through the active participatory efforts of individuals and all interdependent stakeholders. Health is a fundamental human right. We must struggle to provide this right even to those born with unavoidable complications or birth defects.

Health is a natural phenomenon; disease is mostly an invited trouble. Our present system has moved away from nature; right from pregnancy, the newborn is exposed to the risks of falling ill. This could be due to deficient maternal nutrition, the psychological condition of the mother, or environmental pollutants and toxins present in food and water. It could be due to inadequate and unhygienic facilities in the poorer parts of the world. Artificial environments and procedures in modern nursing homes might also be to blame. Knowingly or unknowingly, the seeds of future diseases are sown right from the conception of a new life.

Definition and determinants

The World Health Organization (WHO) defines health as "a state of complete physical, mental, and social well-being and not merely the absence of disease or infirmity." However, in practice, there is no robust and reliable way to measure *well-being*. Generally, the absence of disease or infirmity is considered to be *health*. In many quarters, the terms *health* and *medicine* are used interchangeably, without an appreciation of the difference in the meaning of the two words. Today's health care revolves predominantly around medical care – primarily dealing with disease, diagnosis, and treatments – and not so much on prevention and promotion of health. This mindset has its roots in the misunderstanding of the concept of health. It is necessary to state again: *health* is not synonymous to *medical care*; *health care* is not equivalent to *sick care*, and, as such, it cannot be restricted to encompass only *medical care*.

Many experts feel that the WHO definition of health is not complete without the inclusion of spirituality. The WHO definition highlights the "well-being" of individuals, which is difficult to measure, and hence health care decisions are made based upon the apparent absence of diseases. Moreover, modern definitions of health are mostly restricted to the body and, to some extent, the mind. However, the holistic picture of body/mind/spirit is completely missing in modern descriptions of health. The WHO has been discussing the concept of spiritual health, and several recent studies have endorsed its importance. In 1997, the WHO's executive board resolved to recommend to the General Assembly of the United Nations a new definition of health: "Health is a dynamic state of complete physical, mental, spiritual and social well-being and not merely the absence of disease or infirmity" (cited in Neera Dhar et al).

Four core determinants of health include nutrition, lifestyle, environment, and genetics, which are like four pillars of the foundation of a building. When any one or more of these determinants are compromised, a support system is needed. This is provided by way of medical care, which is considered the fifth determinant (Figure 15.1). Interestingly, two determinants, nutrition and lifestyle, are wholly in our hands, and hence are called

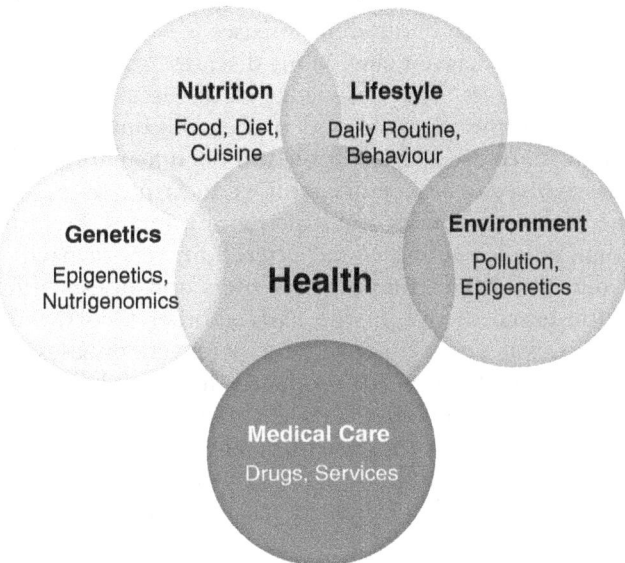

Figure 15.1 Determinants of health

Source: B. Patwardhan, G. Mutalik, and G. Tillu, *Integrative Approaches for Health: Biomedical Research, Ayurveda and Yoga*, London: Academic Press (Elsevier), 2015.

modifiable factors. Many diseases are caused by bad practices of nutrition and lifestyle. The degraded ecosystem and environmental pollution cause several other disorders and diseases. With the help of powerful technology and screening methods, many disorders of genetic origin can be prevented. If one or more core determinants become weak, then only the support of medical care is needed.

Concept of disease

The human body is an auto-regulated, complex system where trillions of cells and molecules continuously interact and grow from conception to death. Different tissues, organs, and systems work in harmony during various stages of development. In a state of health and well-being we are not aware of the existence of any of these complex, dynamic, and continuous processes. Organs and body parts announce their presence only when something goes wrong. It could be a mechanical injury, infection, infestation, or any of a number of other types of assault. These assaults can produce certain symptoms like pain, inflammation, spasm, nausea, vomiting, and diarrhea. This is a natural mechanism which draws our attention to the respective regions of the body and body organs. Certain sets of signs and symptoms are associated with specific diseases. A cluster of specific signs and symptoms can be seen in different diseases.

Knowledge of diseases is not new. Many diseases are mentioned in ancient cultures and civilizations. The knowledge of diseases and the understanding of reasons why the physiological process becomes pathological has evolved through sharp-eyed observation and documentation. This accurate knowledge has evolved over several centuries. In earlier times, diseases were thought to be due to evil spirits and supernatural forces. Prehistoric people practiced rituals, made sacrifices, and uttered prayers to pacify the spirits that caused diseases. Many traditional healers and priests still propagate this superstition-based theory. In the past, another theory, *miasma*, from Greek mythology, was prevalent as to why there were diseases. The miasma theory suggested that diseases are produced due to unhealthy or polluted vapors rising from the ground or from decomposed material. However, as science advanced, faith-based supernatural theories of diseases were challenged. A quest for cause and effect relationships related to signs, symptoms, and diseases gradually developed as the science of pathogenesis and led to diagnostic science.

Pasteur's germ theory and Koch's postulates pinpointed microbial infection as the cause of disease. However, these postulates cannot sufficiently explain the etiopathology of non-communicable diseases. Max Joseph von Pettenkofer (1818–1901) proposed the concept of multi-factorial causation. He linked medicine with physics, chemistry, statistics, and the environment and proposed that diseases may have many causative factors. Later, epidemiology emerged as a new branch of science which addresses the involvement of multiple causative factors.

Diseases like metabolic syndrome and type 2 diabetes are thought to be due to poor nutrition in early life which produces permanent changes in metabolism. This is also known as the thrifty phenotype hypothesis, or Barker hypothesis, which suggests an association of retarded fetal growth to various metabolic diseases in adulthood. This is true for many other serious diseases and conditions of adulthood, such as hypertension, coronary heart disease, stroke, obesity, cancers, and many more. These diseases may be the result of poor nutrition and inadequate care during pregnancy. These are avoidable, undesirable birth gifts, which result in disease, illness, and sickness; there follows an unavoidable sequence of interventions in which medical care and medicines take over the birthright of natural health.

Global health scenario

Health has been a natural and integral part in most of the cultures and traditions. Now, however, peoples' active participation in seeking and maintaining health is overshadowed by their passive dependence on treatments. The basic objective of any health care system is to provide better physical and mental health across communities through effective intervention. Good health care systems also attempt to improve individuals' satisfaction by respecting their dignity. Health is an element of the common good, a central part of economic and social development, and should be accessible and affordable for all individuals.

The world has now crossed a population of approximately 7.2 billion. Out of this, about 80% of the population lives in developing countries. According to the World Health Organization (WHO), about 36 million people have died of HIV so far. Globally, 35.3 million people were living with HIV at the end of 2012. Sub-Saharan Africa remained the most severely affected, with nearly 1 in every 20 adults living with HIV and accounting for 71% of the people living with HIV worldwide. Globally, there are at least 300 million acute cases of malaria each year, resulting in more than a million deaths. Around 90% of these deaths occur in Africa and are mostly young children; malaria is virtually non-existent in the developed world. The number of people who suffer from tuberculosis is over 532 per 100,000 in Africa and Southeast Asia. Spending on pharmaceuticals accounts for about 15% of the total amount spent on health worldwide; the average per capita spending on pharmaceuticals in high-income countries is over four hundred dollars and barely over four dollars in low-income countries.

Over 1 billion people exist on less than one dollar a day. Over 2.5 billion people lack sanitation. Over 1.5 billion people do not have safe drinking water, and some 3 million people – mostly women and children – die every year from diarrheal diseases directly related to these factors. There is little cause for optimism in these areas as available evidence suggests that the drinking water resource in the poor world is likely to diminish over the next several decades.

The epidemiological transition from communicable to non-communicable diseases (NCDs) is apparent in the second decade of the twenty-first century. NCDs were found to be responsible for over 68% of all deaths globally; a near 10% increase from 2000. Earlier in the twentieth century, the rich and developed countries ranked higher in NCDs, while poor countries were contending with communicable diseases. However, over the past few decades, the epidemiological transition is clearly visible where poor and developing countries are confronted with the double burden of both communicable diseases and NCDs.

Successful elimination of many infectious diseases, substantial reduction in childhood mortality, and an improvement of life expectancy and longevity has occurred in the rich and developed world. Ironically, in the poor and developing world, diseases like diphtheria, malaria, and tuberculosis have re-emerged in many regions; new diseases such as HIV/AIDS have become pandemic. Emerging deadly diseases such as Ebola and Zika are immediate threats. Due to the adverse effects of drugs, many new iatrogenic diseases are on the rise. Lifestyle diseases like diabetes are becoming pandemic.

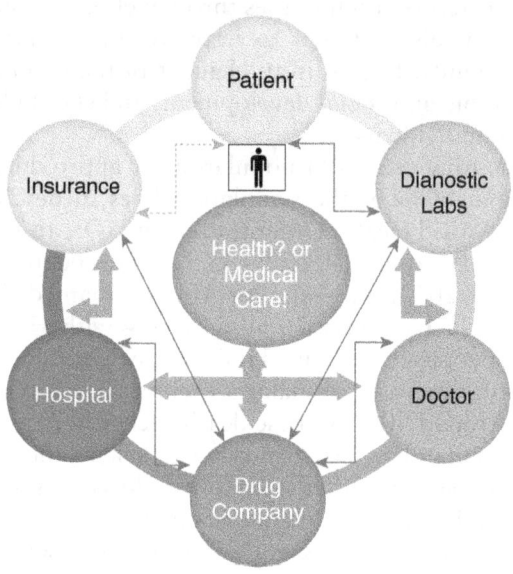

Figure 15.2 A vicious nexus

Source: B. Patwardhan, G. Mutalik, and G. Tillu, *Integrative Approaches for Health: Biomedical Research, Ayurveda and Yoga*, London: Academic Press (Elsevier), 2015.

Sadly, the poor world continues to suffer the ill effects of malnutrition and killer infections like malaria and tuberculosis.

At present, health systems focus more on *sick care* than *health care*. Health seems to be dependent upon medical care. The medical care is dependent upon hospitals. Hospitals are dependent upon doctors. Patients are dependent upon doctors and insurance. Doctors are dependent upon diagnostic tests and products marketed by pharmaceutical companies. Diagnostics are dependent upon pathological laboratories. Pharmaceutical companies are dependent upon the marketing of drugs. Many times, in this vicious cycle of dependency, the interests of people and patients are ignored (Figure 15.2). This situation needs to be considered while discussing the relevance, rationale, and need for integrative approaches to health.

The evolution of the health care system

A good health care system is supposed to deliver quality services to people based upon need. Public health is defined as the science and art of preventing disease, prolonging life, and promoting health through the organized efforts and informed choices of society and people. The WHO's Alma-Ata Declaration is a milestone in the field of public health; it identified primary health

180

care (PHC) as the key to attaining the goal of "health for all" by 2000. While this goal was too ambitious on a global scale, it did help popularize PHC. Besides primary medical care, PHC is focused on basic health determinants such as nutrition, safe water, and sanitation. The drastic reduction of infant and child mortality rates up until the turn of the twenty-first century is certainly attributable to the commendable work done by the WHO at the level of policy-making and by UNICEF at the ground level.

Broadly, the preventive aspects of health care are covered under public health, and the curative aspects are covered under medical care. Health care is provided at various levels. Primary care is the first point where the patient enters the system. This involves preventive care, first aid, and immediate medical assistance. Secondary care is mainly emergency services, and tertiary care is specialized services. Medical care is limited to the professional treatment of illness or injury by providing diagnostic, therapeutic, and surgical services.

Earlier, medical care was more holistic, and the patient was at the center. Clinicians used to treat patients using knowledge, experience, and acumen. In the United States, the Flexner Report of 1910 was a landmark event that gave a big boost to biomedical research through intensive programs aimed at developing high-quality physician-scientists. This opened new avenues for the development of specialties and super-specialties. However, in the process, medical practice became compartmentalized. The earlier clinical practice based on experience and judgment was transformed into a scientific, evidence-based, protocol-driven medical system. As a result, clinical acumen, which is crucial for holistic health, was compromised. Now, the patient, who should be at the center of medical care, is overshadowed by expensive diagnostics; diseases predominate rather than health. The absence of disease is no guarantee of health. Unfortunately, now health care is more focused on drugs, pharmaceuticals, and medical interventions. Optimal health care can and should not be restricted only to drugs or material medicine.

The old health care model based on treatments and therapeutics was more relevant in the twentieth century when the world was facing major threats due to infectious diseases. The domination of non-communicable, chronic, difficult-to-treat diseases is increasing. These diseases are slow-progressing and take a longer time to manifest before leading to life-threatening conditions like cancer. The ageing population is increasing. Therefore, at present, prevention strategies make much better sense than curative approaches. More emphasis on prevention strategies can reduce disease complications and the duration of morbidity before death.

Historically, the medical profession was a noble and service-oriented profession. Providing health care to people was considered the responsibility of the king or the government. Universal health care coverage was the generally accepted norm. In ancient India, the Ayurveda doctor, known as *vaidya*, did not practice for fees. It wasn't common to take fees or money for medical services. This custom changed and private health care providers began charging professional service fees. Over a period of time, this fee-for-service culture became deeply entrenched, and slowly the medical profession became commercial.

According to eminent Indian physician and cardiologist, Dr. B. M. Hegde, "Modern medicine has become a costly chaos with no end in sight. We now have significant problems that beg urgent solutions." Dr. Hegde highlights the need for a new integrated approach in medical science to replace the "present human-made and drug-industry-protected health problems of society can only be physiology-based" (Hegde 2011).

The root cause of many problems in our health care systems of today is related to the commercialization of this sector. The pharmaceutical and diagnostic companies have started defining and dictating health policies. The private service providers began giving better quality health care than the government but at hefty charges. Slowly, the increasing costs of health care necessitated insurance coverage. In reality, the cost of health care became impossible to sustain without insurance coverage. The nexus between insurance companies and private health care providers has taken costs beyond the capacity of common people. This has led to the commoditization of health care and a situation known as the *medicalization of society* (Figure 15.3). Many experts and voluntary organizations are now demanding the *socialization of health care* as a policy.

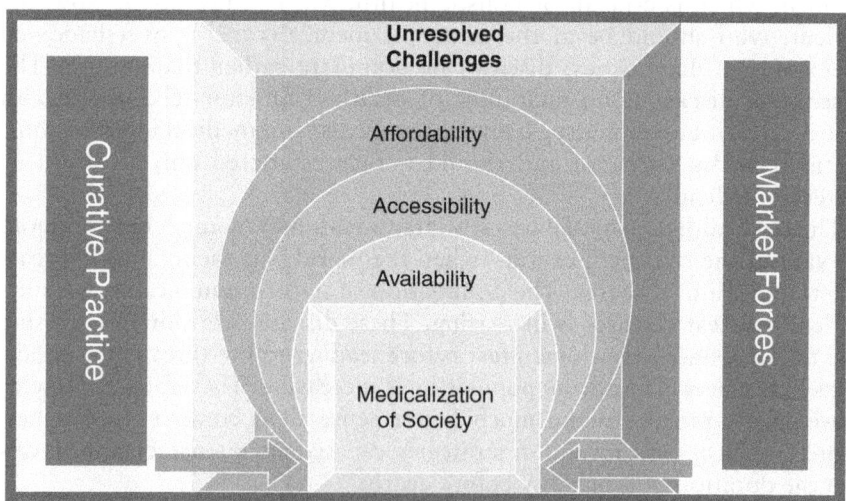

Figure 15.3 Medicalization of society

Source: B. Patwardhan, G. Mutalik, and G. Tillu, *Integrative Approaches for Health: Biomedical Research, Ayurveda and Yoga*, London: Academic Press (Elsevier), 2015.

History of medicine

The history of medicine dates back to the very beginning of human existence. The understanding of health and disease has evolved over thousands of years. Most of the great civilizations – Egyptian, Assyrian, Babylonian, Hebrew, African, Arabic, Chinese, and Indian – had their own traditional systems of medicine. The latter is particularly distinguished in that the tradition which predates the known history of the Sindhu-Sarasvati civilization (one of the oldest world civilizations) has an enduring existence today in the twin disciplines of Ayurveda and Yoga. Reliable archaeological evidence of knowledge and treatments of diseases during prehistoric periods is debatable. The customs and traditions of several African tribes and of the aboriginal people in Australia and the Americas provide a few glimpses. During ancient times, the priests or shamans were responsible for diagnosing and treating diseases.

Prehistoric people, too, knew about medicinal plants and other substances to treat diseases. The *Ebers Papyrus* dating to 1550 BC is considered the oldest and most important medical *papyri* of ancient Egypt. Imhotep, who is thought to be the chief minister to the Egyptian king in the twenty-seventh century BC, is considered to be the first doctor; he is identified with the Greek god Asclepius. Asclepius is the Greek god of medicine, and his daughters represent important branches of medicine. The snake-entwined staff of Asclepius is a symbol of medicine today. Those physicians and attendants who served Asclepius were known as the *Therapeutae* of Asclepius. The use of these terms in modern medicine today shows the influence of these civilizations. There is some archeological evidence that clay tablets bearing specific signs were used by physicians of ancient Mesopotamia. One of the oldest deciphered stone carvings, the Code of Hammurabi, was promulgated by a Babylonian king during the eighteenth century BC and contains a few laws relating to the practice of medicine. Quite surprisingly, the widespread practice of the preservation of dead bodies as mummies by the ancient Egyptians did not lead to a deeper understanding of the human anatomy. However, they contained evidence of arthritis, tuberculosis of the bone, gout, tooth decay, bladder stones, gallstones, and the parasitic disease schistosomiasis – maladies that are seen even today.

Modern medicine was nurtured in the European region and draws substantially from the knowledge and approaches found in *papyri* of Egypt, Greek, and Hebrew origin. While the Harappan civilization also existed during the same time as the ancient Egyptians in the Indian sub-continent, reliable archeological evidence is not available; hence the nature of medical practices prevalent in the Indus Valley civilization is not reliably known. The nature of the art and science of medicine that developed during this period still remains a mystery. Coming to late ancient Christianity one finds that Christian monasteries played an important role in the preservation and development of medical knowledge, since infirmaries were a standard feature of monasteries and convents. The role of monasteries continued in Europe until the Middle Ages. Similarly, in India medical practice was mainly developed by the non-Vedic ascetics like the Buddhists and the Jainas during the fourth and sixth century BC; this is in contrast to the Vedic culture of ritual purity which treated physicians in general as "impure and polluting" (Stuart 2014).

Against this background, it is noteworthy that the Buddhist monasteries played an important role in the practice of medicine and in the development of the classical Indian medical system, Ayurveda (Stuart 2014). According to Schopen, "Indian Buddhist monasteries were well-suited to have infirmaries and provide medical services for the lay community" (Schopen 2004). In contrast to Buddhism's engagement with medicine, that of the other main non-Vedic unorthodox way of life, Jainism, is somewhat ambivalent. As pointed out by many scholars of Jainism, in its canonical phase Jainism did not permit either giving or receiving medical treatment because of their ideas of spiritual progress, based as they are on the cessation of *karmas* by ascetic practices, including extreme bodily discomfort. Though it cannot be the case that the Jainas did not know about the practice of medicine, they did not codify medicine in their monastic tradition (Stuart 2014). But in the post-canonical Jaina texts one finds that "treating the sick in the mendicant order is not only allowed, but in fact, required" (Deo 1952). This shows that there is a shift in the Jaina monastic attitudes toward medicine. The shift is clearly visible in the Shvetambara sect of the Jainas. The Shvetambara texts "make frequent references to medical practice and the alleviation of sickness, describing various medical procedures and instruments and devoting long sections to the interaction between doctors and monastics" (Stuart 2014).

Considering the long, slow process of maturation of thought processes and actual experimentation involved in the making of medicine into a science, as well as the great historical periods in which various schools of thought and ways of life influence each other, it is safe to assume that Vedic, non-Vedic Buddhist, and Jaina traditions were informed and influenced by each other. This fact is important for a comprehensive picture of the history of Indian medicine. However, in the later period, the codification of Ayurveda at the hands of Charaka and Sushruta became the mainstream tradition of Indian medicine. For the purposes of this chapter, the reference point for cause-and-effect-based medicine in India is taken as Charaka's *Charaka Samhita*, the classic of Ayurveda. This classic is supposed to have been written sometime during the fourth century BC. Interestingly, Greek physician Hippocrates' contributions to medicine were also made around this same time.

Indian philosophy

The history of Indian medicine is inseparable from the history of Darshanas, or the schools of Indian philosophy. A detailed account of ancient and contemporary studies on the history of medicine in India has been carried out authentically by Indian scholars. P. V. Sharma's *History of Medicine in India* (1992) and B. V. Subbarayappa's *History of Science, Philosophy and Culture in Indian Civilization*, Vol. 4, Part 2 ("On Medicine and Life Sciences in India") are some notable examples. It has been acknowledged that "medicine was the most important of all the physical sciences which were cultivated in Ancient India" and that medicine was "directly and intimately connected with *Samkhya* and *Vaiseshika* physics" (Dasgupta 273). The various treatises on medicine contain "interesting ethical instructions and reveals a way of life" (Dasgupta 273). As far as the mythological accounts are concerned, the quest for health, wellness, and longevity are manifest in the invocations of Lord Dhanvantari, the incarnation of the supreme deity, Vishnu. The Vedic hymns dedicated to Ashvini Devatas, which are symbolized by the twin stars visible before dawn, represent the healing spirit of the universe.

As regards the systematization of the science and practice of medicine in India, scholars agree that the history of Indian medicine is synonymous with Ayurveda, which is supposed to be a part of Atharvaveda. About its antiquity Charaka says, "there were always medicines which acted on human body according to the principles of Ayurveda" (Dasgupta 274). However, the status of Ayurveda is a contested matter. For Charaka, it is a full-fledged Veda and is in fact superior to the other four Vedas, and therefore it is called the fifth (*panchama*) Veda. For Vagbhatta, it is an *upa-veda* (independent division) of Atharvaveda, while for Sushuruta it is an *upanga* (part) of Atharvaveda.

All this shows that Atharvaveda in some way existed when the Vedic literature was still in the making. Along with Atharvaveda, one finds references to other compendiums of medicine, such as *Subhesja* (Dasgupta 276). It is also interesting to note "nowhere in the Upanishads or Vedas does the name *Ayurveda* occur, though different branches of study are mentioned in the former" (Dasgupta 276).

As far as Indian philosophy and the history of medicine in India is concerned, it should be noted that along with Ayurveda, which belongs to the Vedic tradition, non-Vedic Buddhism, as a philosophical creed and as a way of life, has significantly contributed to Indian medicine. One of the earliest systematic mention of medicines "unmixed with incantations is found in the *Maha-vagga* of the *Vinaya-Pitaka* where Buddha prescribes medicine to his disciples. These medicines are of a simple nature, but they bear undeniable marks of methodical arrangements" (Dasgupta 276). The *Vinaya-Pitaka* contains special sections on specifications about the food and medicine of Buddhist monks and classic Buddhist descriptions of meditation on the body with lists of body parts and bodily impurities. For example, in texts such as *Bhaishajyavastu* of the *Mulasarvastivada* and the *Vinaya-Pitaka*, there is a detailed description of medication allowed for Buddhist monks and nuns (Stuart 2014).

There is a mention of a surgeon, Akashagotto, who made surgical operations (*sattha kamma*) on fistulas. There is a mention of Jivaka, who studied medicine in the Taxila University under Atreya (Stuart 2014). As pointed out earlier, this shows that the practice of medicine was common even among the non-Vedic traditions, such as the Buddhist and Jaina traditions, even at the time of Atharvaveda. Scholars have debated the issue of the antiquity of Atharvaveda and how Ayurveda is related to it. The two main codifications, Charaka's *Charaka Samhita*, and Sushruta's *Sushruta Samhita*, allude to the traditional continuity of Ayurveda. In this context, Ayurveda means "life, the constancy of the qualities of medical herbs, diet, etc., and their effects on the human body, and the intelligent inquirer" (Dasgupta 277).

Although Samaveda and Atharvaveda refer to health, disease, medicine, poisons, and spiritual and remedial measures, the Vedas, as such, are generic in nature and do not contain any particular metaphysical discipline of thought. The systematization of different metaphysical views resulting in various Darshanas[2] like Samkhya Yoga, Mimamsa, and Vedanta is a latter phenomenon. Darshanas are broadly classified into Astika, who believe in *atma*, and Nastika, who do not. Vedanta is one of the Astika Darshanas. Purva-Mimamsa (ritualistic view) and Uttara-Mimamsa (advaita) have different approaches toward the Vedas.

In this process, the Vedanta, as a philosophical school, is a culmination of the philosophical views of Samkhya and Uttara-Mimamsa. Ayurveda derives its basic premises from the Samkhya Yoga as a combined discipline; the other premises were derived from the Vaisheshika and Nyāya systems. These philosophical theories play a fundamental role in Ayurveda in the examination of the physical, mental, social, and spiritual health of an individual. This function is encompassed in the term *the science of life.*

Among the Darshanas, it is apparent that the Samkhya system had the greatest influence on the precepts of Ayurveda. The aim of the Samkhya system is the cessation of sufferings (*dukkha nivritti*). This system is attributed to the legendary sage Kapila. It aims at comprehending the entirety of the meaning of life through twenty-five *tattva* (principles). It is interesting to note that the Samkhya begins its inquiry into the good life by identifying and distinguishing between three types of *dukkha* (suffering). The three types of suffering are *adhibhuatika* (caused by external conditions, mainly physical), *adhidaivika* (caused by supernatural agency), and *adhyatmika* (intraorganic). The twenty-five *tattvas* are the principles of the creation of the universe; the five fundamental derivatives of the universe, *pancha mahabhutas* (earth, water, energy, air, space), and the *tri-gunas* (the three primordial natures of *prakriti* coded as *sattva*, *rajas*, and *tamas*) are at the foundation of the evolving philosophy of Samkhya. These were translated and applied to all the problems of health and wellness in Ayurveda. The most fundamental derivation of this is the conceptualization of the *tri-doshas* as *vata*, *pitta*, and *kapha*, which are responsible for all kinds of morbidity in the body.

Evolution of Ayurveda and other disciplines

Notable contributions to the disciplinarization of Ayurveda can be traced to the pioneering work of Charaka, the father of Indian medicine. The *Charaka Samhita* elaborates principles, diagnoses, and treatments for diseases. This classic text gives a detailed knowledge of the human anatomy, embryology, pharmacology, blood circulation, and diseases such as diabetes, tuberculosis, heart disease, as well as details of thousands of medicinal plants used in Ayurveda. In other scientific disciplines like astronomy and mathematics, significant contributions followed Charaka, including the work of the master astronomer and mathematician Aryabhatta (476 AD). In his classic text *Aryabhatiya*, he describes the process of measuring the motion of planets and eclipses. Aryabhatta proclaimed that the earth is round, that it rotates on its axis, and that it orbits the sun and is suspended in space. Aryabhatta's most significant and well-known contribution is the concept of zero. Varahamihira (499–587 AD), in a classic text, *Pancha Siddhanta*, notes that the moon and the planets are lustrous not because of their own light but due to sunlight. In the *Bruhad Samhita*, he details many discoveries in the domains of geography, botany, and animal science. Another great scholar and master of arithmetic and astronomy was Bhaskaracharya (1114–1183 AD). In his classic text *Surya Siddhanta*, he made a note on the force of gravity. Nāgārjuna (800 AD), in the classic text *Rasa Ratnakara*, outlined many interesting experiments in metallurgy, alchemy, and chemistry. He also used metal-based ingredients in medicinal preparations.

After Charaka, there were only two major contributions in the next ten centuries: Vagbhata (circa sixth to seventh century AD) wrote two treatises, *Ashtanga Hridaya* and *Ashtanga Sangraha*; and the *Madhava Nidana* (circa ninth century AD), an Ayurvedic text on symptomatology and diagnosis. Indian science left its footprint in neighboring regions: from 400 BC, Indian knowledge became available to European and Arabic regions through invasions by Alexander the Great and others. There is evidence of some communication between Greece and India even before Alexander's invasion, possibly through Persians of the Achaemenid Empire (Filliozat 1964).

A closer look at the history of Western science reveals the influence of Vedic and ancient Indian knowledge systems on many philosophers and thinkers from the West. Evidence of this influence is also found in the works of pioneers in the field of medicine such as Hippocrates, Avicenna, Galen, and Osler. The influence of Vedic knowledge on Greek science and medicine is apparent, yet hardly any mention or recognition appears in the documents, including in Osler's often-cited book on the history of modern medicine (Osler 233). Whether Osler's omission of Ayurveda in his landmark book is due to his ignorance or a case of biased negligence is a subject of speculation (Patwardhan et al. 2003).

Whereas most of Western scientific, philosophical, and medical texts do not acknowledge the influence of other cultures, the Persian scholars duly acknowledged their sources of knowledge. For instance, Abdulla-bin-Ali, Manka, and Ibun-Dhan translated several Ayurveda texts into Persian and Arabic. *Charaka Samhita* (Sharaka), *Sushruta Samhita* (Kitab-E-Sushrud), *Ashtanga Hridaya* (Astankar), and *Siddhayoga* (Sindhashtaq) reached the Arabic and Persian physicians in the sixth to eighth centuries. In 850 AD, Ali-bin-Raban-al-Tabri authored a textbook of medicine, *Firdausu'l-Hikmat*. He appended a chapter on the Indian system of medicine in this book. This was much before Avicenna wrote his famous *Canon of Medicine*.

The process of dialogue, dialectical method of inquiry, and hypothesis building is common to both Indian and Western philosophy. The Delphi method of participatory consensus building, which is used widely today, resembles the method in Ayurveda known as *Tadvidya Sambhasha*. Detailed rules and procedures for conducting consultations, debates, and discussions are given in the Nyāya Shastra which as a method of argumentation was common to all disciplines including Ayurveda. The Greek theory of macro-cosm and microcosm, the concept of the four elements (fire, air, water, and earth), the four qualities (hot, dry, wet, and cold), and the four humours (yellow bile, blood, phlegm, and black bile) are also to be found in the ancient systems of Samkhya and Yoga. These formed the basis of *Loka-Purusha*, the five *Mahabhoota*, three *Dosha*, and six *Rasas* which form the basis of Ayurveda.

The Indus Valley Civilization, and the ancient Chinese, Mesopotamian, and Egyptian civilizations share several similarities. In the beginning of the first millennium AD, three principal systems of medicine, namely Ayurveda, Greek, and Chinese, were active. The fundamental attitude to the relationship of man and nature was similar in these three ancient civilizations. However, their explanations of the human body, physiology, pathology, and therapy were different in some respects (Subbarayappa 2001). Their approach to health and medicine was holistic. Later, Greek medicine followed the modern science path and gradually evolved as modern medicine. Ayurveda emerged as a remarkable holistic system, both in its foundational ideas and therapeutic measures. Later, during the period of foreign invasions in India, Ayurveda was suppressed and stagnated for several centuries, dealing a severe setback for the systematic progress and development of Ayurveda. Moreover, because of the dominance of modern medicine during the colonial period, Ayurveda was further fragmented, deteriorated, and marginalized.

Philosophy and science

During the ancient times science and philosophy were considered as one whether in the West or the East. The ancient Egyptian, Roman, and Greek thought traditions took a holistic approach to man, nature, and the universe. However, these holistic traditions were slowly replaced by modern science by its scientific method of observation, rigorous experimentation, and analysis. Some of the well-known philosophers and scientists who represent the spirit of modern science are Galileo, Descartes, Pascal, Newton, and Einstein. As modern science was slowly replacing the ancient Greek scientific outlook that was mainly shaped by Aristotle, many of the ancient Eastern traditional knowledge systems were becoming frozen parts of history for a variety of reasons including social, political, and economic. They remained static for a long time and lost the dynamism implicit in the spirit of experimentation that is necessary for the progression of science. As a result, many of the Eastern traditions, like that of Tibet, remained more a philosophical bulwark against the adaptation of the culture of science. The ancient Indian knowledge systems like Ayurveda and Yoga seem to fall somewhere in between.

Western philosophical tradition is mainly divided into two main standpoints: rationalism and empiricism. As the very term suggests, rationalism stresses reason as the chief source of knowledge of the universe, that is, human reason is objective and is the chief source of validity of human experience. Against this, empiricism holds that the main source of knowledge is sensory experience. The logic that was developed by Aristotle takes reason as the source of valid reasoning in which particular conclusions are drawn from the general premise by way of the process of deduction. This is called a deductive logic and is used mainly in formal disciplines like mathematics and logic.

On the other hand, the logic that was developed by empiricist philosophers such as David Hume is based on the grounds of experience which can be generalized; that is, a general conclusion is drawn from observing several particular facts. Until the Middle Ages, Aristotelian deductive logic was rigorously used in understanding several natural phenomena and in their explanations. But Aristotelian logic as a tool of scientific reasoning and as a means of classification of causes was inadequate to explain many natural phenomena discovered during the Renaissance. A new scheme of explanation of natural phenomena was needed. In this new scheme of explanation there was no place for any supernatural power, or God, or scriptural authority. The whole universe was governed by a system of natural laws and man was supposed to *discover* these laws. There was no purpose to the universe. The universe was governed by purely mechanistic laws. It was not governed by the will of God. In this universe, man was at the center of all inquiry. This was the spirit of the Renaissance that gradually prepared the way for the evolution of modern science. However, in the process of scientific advancement, valuable insights and ancient knowledge were neglected, lost, or remained in obscurity.

Philosophy of medicine

Philosophical questions about any discipline constitute a meta-discipline. Thus, there are such disciplines as the philosophy of mathematics, the philosophy of history, or even the philosophy of music, wherein philosophical questions are raised about various concepts and procedures in mathematics, or history, or music. One such overarching meta-discipline is the philosophy of science. As far as the philosophy of medicine is concerned, there are diverse views regarding what constitutes a philosophy of medicine. In 1992, one of the authorities in this field, Arthur Caplan, raised the question, "Does the philosophy of medicine exist?" According to Caplan, the philosophy of medicine is a sub-discipline of the philosophy of science (Caplan 1992). The philosophy of medicine includes medical ethics and bioethics, but its scope is much wider, involving epistemological, metaphysical, and methodological dimensions of medicine, which include therapeutic, experimental, diagnostic, and palliative aspects. If we consider medicine only as a branch of the basic sciences like anatomy, physiology, and biochemistry, then the philosophy of medicine will be a branch of philosophy of science.

In reality, medicine is not limited to patho-physiology, pharmacology, psychology, microbiology, or genetics. The main purpose of the science of medicine is to understand human body functions and dysfunctions through sound hypothesis, robust experimentations, observations, analyses, and a continued search for truth. However, the real purpose of medicine is more than this. Indeed, it is a search for truth, with the definite objective to attain health and healing. Probably this is the reason why religious places like churches, temples, monasteries, and mosques were involved in medical care before the evolution of clinics and hospitals. These sacred places protected traditional knowledge during invasions and dark periods in their respective regions. In many ancient civilizations like Egypt, priests worked as doctors and healers. Even now, in some parts of Africa and India, such practices exist. In ancient times, medicine was not restricted to material drugs; rather, the approach was holistic, involving body, mind, and spirit.

The East and West: the differing attitudes

Before the European Renaissance, the division of the world into East and West was not very sharp. The European Renaissance of the fourteenth to the seventeenth centuries AD started in Italy and spread throughout Europe. This triggered a new era of creation, transformation, and progress and witnessed spectacular revolutions in humanistic, social, political, and intellectual pursuits. Some of the world's greatest artistic achievements were made during this period. Remarkable artists like Leonardo da Vinci and Michelangelo made timeless impacts during this period. The fruits of the Renaissance in various disciplines including art, culture, science, technology, medicine, and the humanities were mainly seen in the West, the Middle East, and Russian regions. Although many of the technological innovations reached the East through European travelers and merchants, the *spirit* behind these developments in science, technology, and humanities was not captured by the

Easterners, including the Indians. As a result, today's modern science is known as Western science and modern medicine is known as Western biomedicine.

Western science and philosophy is a mix of many strands which include in principle the rationalist strand held by philosophers like Aristotle, Descartes, Spinoza, Leibniz, and Kant and the empiricist strand held by John Locke and David Hume. The history of Western science is closely associated with the developments in the rationalist and the empiricist schools, since these are the two basic epistemological positions held by all major philosophers and scientists in the West. The whole history of Western philosophy and science has shaped the modern world through robust developments in various branches of sciences and technology. In the light of these developments, the modern world is characterized as more rational, logical, mechanistic, materialistic, and human-centered. On the other hand, the Eastern world continued with its traditions and has remained suppressed, intuitive, and spiritual. If an epistemological bridge, a two-way traffic, had existed between these two worlds, the face of Ayurveda, as well as modern medicine, would have been much different.

Often it is said that the Eastern and the Western ways of thinking about man and the universe differ in many ways. The Western mind and philosophy is alleged to be more restrictive (for example, to sensory organs), reductive (for example, to mind and matter), quantitative, and analytical – relying more on empirical, measurable parameters within the understanding or experiences of human sensory organs, or from instruments as their extensions. On the other hand, the Eastern mind and philosophy, particularly the Indian, is alleged to be inclusive, holistic, qualitative, and intuitive in its approach, not restricted to sensory experiences, and often aspiring to transcendence of the body/mind. These differences are articulated and theorized variously by Indologists and sociologists. What is relevant to note is that these differences were articulated against the background of modern Europe's self-perception vis-à-vis the East, or the Orient. Thus, the articulation of these differences has a definite political background involving the complex relationship between Europe and the Orient. This is clearly reflected in the works of German Romantics like Schlegel and Herder. India, in their imagination, is the land of *spirit*, whereas, as in the imagination of Hegel, Europe is the land of *reason*.

Another dimension of these characterizations of the East and the West, particularly Europe and India, is clearly political and sociological. It has been argued that characterizations such as rational versus emotional are grounded in the deep-seated bias that modern Western civilization has toward the East. This bias divides the social reality into two realms, "the realm of this side of the line" and the realm of "the other side of the line." This division is based on granting the highest rationality to modern science and technology and denying the status of rationality to other forms of non-Western thinking and philosophy (Deshpande 5). Such distinctions are often introduced as analogous to brain structure and functioning. If the Earth can be visualized as a human brain, then one can say, metaphorically, that

191

the philosophies and sciences of the East and South are dominated by the right brain and that of the West and North are dominated by the left brain. Interestingly, the difference in Eastern and Western approaches is seen in the respective geographical regions as well.

We are aware that while psychologists agree with the left and right brain theory, brain scientists differ. We feel that psychology, as a systemic science focusing on the mind, represents the right side, while neurology, as a structural science focusing on the brain, represents the left side. According to Jeff Anderson, a neuroscientist from the University of Utah who studied over seven thousand regions of the brain with the help of magnetic resonance imaging (MRI):

> It's absolutely true that some brain functions occur in one or the other side of the brain. Language tends to be on the left, attention more on the right. But people don't tend to have a stronger left- or right-sided brain network.
>
> (Nielsen 2013)

At a physical level, the division between the left and right brain may not be seen by MRI. A brief history of ideas about the brain, philosophical approaches, and contemporary discussion on consciousness from a neuroscience perspective is very interesting (Chakravarti 428). Recently, researchers have discussed the complex relationships between the hemispheres from evolutionary and sociological perspectives (Michael 2011).

We feel that modern science still has many limitations in understanding the real functioning of the brain, let alone its relationship to an entity called *mind*. The analysis of the brain by neuroscientists is a typical feature of the Western, analytical mind; an artist's depiction of the brain as a globe, showing the right and left divide, is a typical feature of the Eastern, creative mind. We are considering the division of East and West not merely in a geographic sense but more in terms of approaches and mindsets. Both the Eastern and Western approaches are important, and should not be seen as mutually exclusive. Medicine has a much larger role than that of physical science. It is not merely a sum of philosophies of science, biology, or the humanities. The motivating force of the practice of medicine is the alleviation of suffering and restoration of people's health (Velanovich 1994).

In search of roots

In this brief philosophical and historical overview, I have highlighted that Indian systems of knowledge, including medicine, were not only hamstrung by invasions and foreign rule but were denied revitalization through the influence of the industrial revolution and the progress of science and technology that propelled Western science to prominence. The important question at this juncture is how to rediscover our roots and restore growth, development, and vitality to the traditional science systems in India – mainly to Ayurveda and Yoga. Yoga itself has undergone a global transformation through multiple channels. Yoga has spread globally through Buddhism, including Western institutes of excellence where through scientific research and scrutiny it has gained credibility and wider acceptance.

Today we find that centers of excellence in medicine such as Mayo Clinic, Cleveland Clinic, and other prominent academic centers of medicine bring Yoga practices into rehabilitative medicine and other fields of physical and mental health. In other channels in Europe and the United States, where through individual initiatives of Indian scholars like Maharshi Mahesh Yogi (the originator of transcendental meditation), many types of Yoga such as mindfulness and Hatha Yoga are practiced. Yoga has also been lamentably commercialized and often reduced in modern Western cities to street-corner outlets purveying quick fix, body-building exercises. The challenge is to bring to the West the quintessential feature of Yoga – based on an inextricable and indivisible meld between body and mind – and apply it as the basis of a way of life for the prevention of illness, promotion of health, restoration of vitality, and treatment of diseases.

Ayurveda, according to some scholars, cannot be called a strict science. At the same time, Ayurveda does not oppose science and is not unscientific or pseudo-scientific. Ayurveda is considered a trans-disciplinary knowledge system that accepts but transcends methods of science (Manohar 2005). Along with integrating its practices with yogic disciplines and modern medicine, Ayurveda has the potential to emerge as a major scientific discipline. Before that happens, there are many hurdles to clear.

The first hurdle is establishing an acceptable evidence base suited epistemologically to Ayurvedic precepts and practices. Secondly, efforts to identify an effective therapy in Ayurveda, discover herbal remedies, or isolate the active principles which have maximum therapeutic properties – as is the current approach to Ayurveda – is not the right way to go. This may yield a few short-lived therapeutic agents and not the revitalization of Ayurveda as a whole since this is purely a reductive methodology. Thirdly, Ayurveda must develop a bridge through a "systems biology approach" through the scientific characterization of well-linked components at various levels of application of cell biology, biochemistry, genomics, epigenetics, proteomics, and metabolomics (Patwardhan 2014). Moreover, basic Ayurvedic principles, such as *Mahabhuta*, *Tridosha*, and *Prakriti*, and their contribution to pathophysiology need to be scientifically explored with tools of advanced technology. Fourthly, training in Ayurveda is today confined to a class of students who are filtered down after the best are absorbed in various careers of business, technology, modern medicine, and other monetarily rewarding careers. Unless this is changed, it will be difficult to get the brightest students to emerge as leaders of Ayurveda of the new era. Educationists, thought leaders, and policymakers need to apply their minds seriously to this daunting challenge. Fifthly, in the interim, there is a need to train scientists as *vaidyas*, as yogis, and as modern scientists in order to create a confluence of traditional wisdom and modern science (Patwardhan 2011). True success will be the manifestation of new models of integrative health and medicine to benefit the global community – with Ayurveda and Yoga as its principal ingredients (Patwardhan 2015, 353).

Broader vision of integration

The value of integrative approaches is becoming increasingly clear in many super-specialty areas such as oncology, cardiology, neurology, dermatology, psychiatry, and geriatrics. Modern medicine alone cannot fulfil our present requirements. Biomedical professionals should not monopolize medical care or health care. The limitations of modern medicine – especially in managing chronic, behavioural, and lifestyle diseases – are becoming clearer. The present approach of evidence-based medicine must not become too rigid and restrictively protocol-based. Modern medicine cannot lose sight of the person behind the patient. At the same time, Ayurveda and Yoga professionals cannot continue to position themselves as practitioners of ancient traditional systems anymore. They can neither exist merely on pride and past glory nor remain ritualistic. They must emulate modern science. They must be open to questioning and experimentation. They may be proud of their experiential heritage but must also respect the value of experimental evidence. It is high time to discard their silo mentality and uphold the public good at the top.

There is an urgent need to integrate modern and traditional systems. Practitioners' mindsets should be marked by mutual respect, accepting the strengths and limitations of their own medical systems. Practitioners from different spheres need to collaborate and work together instead of being at loggerheads. It is heartening to note that the concept of integrative medicine is now getting wider recognition from credible scientific bodies like the Institute of Medicine (IOM) of the National Academies in the US. A preface in the landmark IOM report on integrative medicine is indicative of future trend (Schultz et al. 2009. Undoubtedly, there is growing consensus among top medical professionals and establishments that an integrative approach is simple, scientific, affordable, and patient-centric, which can transform the current health care system into a vibrant, evidence-based, holistic, and humane health care system of the future.

Principles for integration

True integration is not to be mistaken for merely cross-practice or a pluralistic system of medicine. The integrative approach is required at many levels (Figure 15.4).

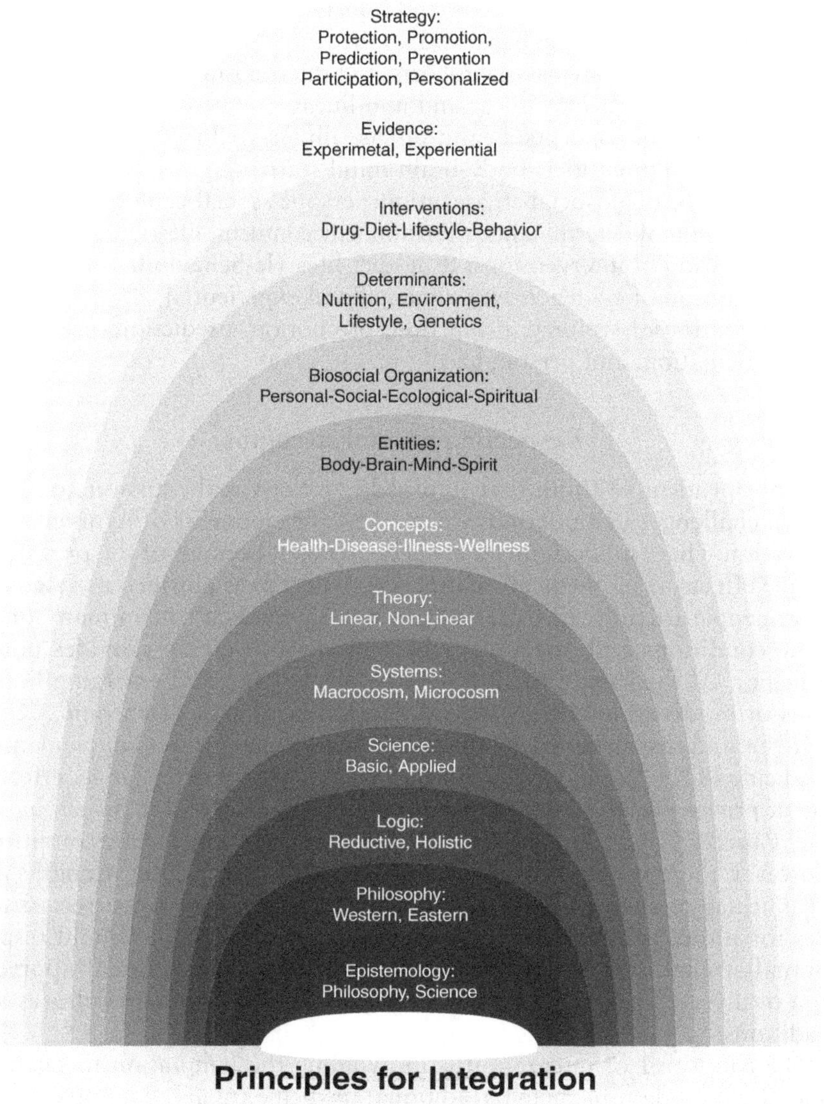

Strategy:
Protection, Promotion,
Prediction, Prevention
Participation, Personalized

Evidence:
Experimetal, Experiential

Interventions:
Drug-Diet-Lifestyle-Behavior

Determinants:
Nutrition, Environment,
Lifestyle, Genetics

Biosocial Organization:
Personal-Social-Ecological-Spiritual

Entities:
Body-Brain-Mind-Spirit

Concepts:
Health-Disease-Illness-Wellness

Theory:
Linear, Non-Linear

Systems:
Macrocosm, Microcosm

Science:
Basic, Applied

Logic:
Reductive, Holistic

Philosophy:
Western, Eastern

Epistemology:
Philosophy, Science

Principles for Integration

Figure 15.4 Layers of integrations

Source: B. Patwardhan, G. Mutalik, and G. Tillu, *Integrative Approaches for Health: Biomedical Research, Ayurveda and Yoga*, London: Academic Press (Elsevier), 2015.

- Integration of epistemology: philosophy and science
- Integration of philosophy: Western and Eastern
- Integration of logic: reductive and holistic
- Integration of science: basic and applied
- Integration of systems: macrocosm and microcosm
- Integration of theory: linear and non-linear
- Integration of concepts: health-disease-illness-wellness
- Integration of entities: body-brain-mind-spirit
- Integration of biosocial organization: personal-social-ecological-spiritual
- Integration of determinants: nutrition, environment, lifestyle, and genetics
- Integration of interventions: drug-diet-lifestyle-behaviour
- Integration of evidence: experimental and experiential
- Integration of strategy: protection, promotion, prediction, prevention, participation, and personalized

Respecting mutual strengths

It is not prudent to think that Ayurveda or Yoga is the answer to all the unmet challenges in modern medicine. Many proponents of Ayurveda take a position where modern medicine is disparaged because of the prevalence of side effects of biomedical treatments; Ayurveda is glorified as safer, and more people-friendly. In fact, it might not have much role in many of the acute conditions and infections where modern medicine provides potent remedies. At the same time, because of the existing and emerging limitations of modern medicine, especially in the treatment of chronic, degenerative, and psychosomatic diseases, Traditional and Complementary Medicine (T&CM) may have a larger role to play. For instance, effective use of preventive and immunity-building measures through *Swasthavritta* and *Panchakarma* can reduce the incidence of many acute conditions. However, there is a need to do a critical reassessment of the strengths and weaknesses of each of these approaches; this will lead to correct strategies for future medical therapy and health care. While we should respect mutual strengths, it is equally important to develop a mutual appreciation of diverse approaches, views, and philosophies from all cultures and traditions.

The true spirit of integration is in avoiding the temptation to take any sides, be it modern medicine, traditional medicine, Ayurveda, Yoga, or any other system. True integration is about a scientific and unbiased attempt to strike a mutual trust-based balance between the various systems in the best interest of the people. It is high time modern medicine and traditional medicine, including Ayurveda and Yoga, seamlessly integrate to emerge as the mainstream medicine of tomorrow. Future doctors need to become more human, humble, and humane in their approach to patients in distress. They need to give priority to patient care rather than hubristic promotion of any particular system, be it modern medicine or traditional medicine. Also, the compassion and care shown by a good doctor is as important as any existing medicines or therapies. The sooner a roadmap emerges to hasten the process of integration the better it will be to heal today's sick planet. Besides curing illness, the quintessential mission of medicine is to prevent disease, promote health, restore vigor, and help ensure the highest quality of life at every stage of life. This is a prodigious dream and an overarching aspiration of humanity; it is not going to be realized easily or spontaneously. The entire global medical community – practitioners, scientists, and specialists – should meet this grand challenge and embrace integrative approaches to health care, thus furthering the goal of health for all.

Notes

1 Portions of this chapter were previously published in B. Patwardhan, S. Deshpande, G. Tillu, and G. Mutalik, "In Search of Roots: Tracing the History and Philosophy of Indian Medicine," *Indian Journal of History of Science*. Vol. 50.4 (2015): 629–641; and B. Patwardhan, G. Mutalik, and G. Tillu, *Integrative Approaches for Health: Biomedical Research, Ayurveda and Yoga* (London: Academic Press, Elsevier, 2015).
2 The term "darshana" literally means *to see. Drushyate anena iti darshanam*: Darshana is the one which enables one to perceive deeply.

Works cited

Caplan A. "Does the Philosophy of Medicine Exist?" *Theoretical Medicine*. Vol.13.1 (1992): 67–77.

Chakravarti, S.V. "Demystifying Brain." National Program on Technology Enabled Learning (NPTEL), Government of India, IIT Madras, India. http://nptel.ac.in/demystifying.php.

Dasgupta, S.N. *A History of Indian Philosophy*. Vols. 1–5. New Delhi: Motilal Banarsidass Publishers Pvt Ltd. (Original work published 1922).

Deo, S.B. "The History of Jaina Monachism from Inscriptions and Literature." *Bulletin of the Deccan College Research Institute*. Vol.XVI (1954–55): 1–4.

Deo, S.B. *History of Jaina Monachism from Inscriptions and Literature*. Mumbai: Univesity of Bombay, 1952.

Deshpande, Sharad. (ed.). *Philosophy in Colonial India*. New Delhi: Springer, 2015.

Dhar, Neera, S.K. Chaturvedi, and Deoki Nandan. "Spiritual Health Scale 2011: Defining and Measuring 4th Dimension of Health." *Indian Journal of Community Medicine*. Vol.36.4 (October–December 2011): 275–282. www.ncbi.nlm.nih.gov/pmc/articles/PMC3263147/.

Filliozat, J., and Dev Raj Chanana. *The Classical Doctrines of Indian Medicine*. New Delhi: Munshiram Manoharlal Oriental Booksellers and Publishers, 1964. 229–257.

Hegde, B.M. "Integrated medical care system (complementary systems of medicine – are they scientific?)." *Journal of Indian Academy of Clinical Medicine* 12(4) (2011): 260–262.

Manohar, R. "Ayurveda as a Knowledge System." *Indian Knowledge Systems*. Vol. 1. Eds. Kapil Kapoor and A.K. Singh. Shimla: Indian Institute of Advanced Study; New Delhi: DK Printworld, 2005. 156–171.

Michael, T. "The Master and His Emissary: The Divided Brain and the Making of the Western World." *Cognitive Neuropsychiatry*. Vol.16.3 (2011): 284–288.

Nielsen, J.A., B.A. Zielinski, M.A. Erguson, J.E. Lainhart, and J.S. Anderson. "An Evaluation of the Left-Brain vs. Right-Brain Hypothesis with Resting State Functional Connectivity Magnetic Resonance Imaging." *PLoS One*. Vol.8.8 (2013): e71275. doi:10.1371/journal.pone.0071275.

Osler, S.W. *The Evolution of Modern Medicine*. Ed. F.H. Garrison. Silliman Memorial Lectures. New Haven: Yale University Press, University of Virginia Library Electronic Text Center, 1921. 233.

Patwardhan, B. "Bridging Ayurveda with Evidence-Based Scientific Approaches in Medicine." *EPMA Journal*. Vol.5.19 (2014): 1–7. doi:10.1186/1878-5085-5-19.

Patwardhan, B., A. Chopra, and A. Vaidya. "Herbal Remedies and the Bias Against Ayurveda." *Current Science*. Vol.84.9 (2003): 1165–1166.

Patwardhan, B., V. Joglekar, N. Pathak, and A. Vaidya. "Vaidya-Scientists: Catalysing Ayurveda Renaissance." *Current Science*. Vol.100.4 (2011): 476–483.

Patwardhan, B., G. Mutalik, and G. Tillu. *Integrative Approaches for Health: Biomedical Research, Ayurveda and Yoga*. 1st ed. San Diego: Academic Press, 2015.

Schopen, Gregory. "The Good Monk and His Money in a Buddhist Monasticism of the Mahayana Period." *Buddhist Monks and Business Matters: Still More Papers on Monastic Buddhism in India*. Honolulu: University of Hawaii Press, 2004. 1–18.

Schultz, A.M., S.M. Chao, and J.M. McGinnis. *Integrative Medicine and the Health of the Public: A Summary of the February 2009 Summit*, Washington, DC: National Academies Press, 2009, 222pp.

Stuart, M.J.. "Mendicants and Medicine: Āyurveda in Jain Monastic Texts." *History of Science in South Asia*. Vol.2 (2014): 63–100. http://hssa.sayahna.org.

Subbarayappa, B.V. *On Medicine and Life Sciences in India*. History of Science, Philosophy and Culture in Indian Civilization, 1st ed. Vol. 4, Part 2, New Delhi, India: Center for Studies ion Civilization, 2001. 771pp.

Velanovich, V. "Does the Philosophy of Medicine Exist? A Commentary on Caplan." *Theoretical Medicine*. Vol.15 (1994): 77–81. doi:10.1007/BF00999221.

16

A CONNECTICUT YANKEE IN INDIRA'S COURT

A brief account of modern Indian science

Raman Srinivasan

Rain clouds

Washington, D.C, 1966

When the dark blue monsoon clouds fail to arrive, you are in trouble. In fact, the whole Indian subcontinent is in trouble. The harvest will be erratic; later, rice will become scarce. Then prices shoot up, like Fourth of July fireworks. People will starve. "In the great famine they ate grains," some British historian might have told you, "culled from cattle dung." And so you fear that if India is in trouble, then the communists, both Russians and Chinese, may take advantage of the famished. Their sphere of influence will grow. You are certain he does not want the pivot of the world turning red. So you send weekly reports on the monsoons in India to LBJ. He likes the colourful maps.[1]

New Delhi, 1966

One can either pray to the gods or the Americans for the monsoons. Who, just who, is to be invoked? You know that it is the duty of the monarch (and especially so in a parliamentary democracy) to ensure that the rains are not only abundant but also timely. Of course, in the past, you, the only daughter of the Noble Nehru, have asked friendly Americans to bring uplifting undergarments. Later, those ingenious Americans even helped re-sculpt your ingressive Nehru nose. But then they also desired that you put your new beak in unmentionable places. In any case, the god of waters, Varuna, lives in the West, and maybe these American druids do know something hitherto unknown.[2]

Chattarpur, 1966

However, if you never believed those engrossing Americans, and counsel caution to the cabinet secretary of Mrs. Indira Gandhi, you find yourself transferred to a semi-arid scrub-jungle, even if you be the mighty director-general of the Department of Meteorology of the Union Government of India. So you adhere to the Truth, and leave three bright-eyed Central School-going children and a dutiful wife behind in Delhi, and go live the life of a bachelor-ascetic in Bundelkhand. And you find out that every cloud does have a silver lining, even the cloud of unknowing. Thanks to the Americans sprinkling silver salts over the pre-monsoon clouds, you meet the living Truth, and therefore sing: *He is not only near, he is verily here.*

He answers your thousand questions unasked in a dilapidated government bungalow in Bundelkhand. The sun rising in Bundelkhand is like a delicious golden ball, a levitating *laddu*, and He, your patient teacher, makes you comprehend, as it were, the meaning of meaning:

> Through the fuel of the Absolute is Truth
> proffered wholly to the Ignition in fire;
> recognizing this in all action one reaches That.[3]

Sir, wise teacher, why are the rain-bearing clouds erratic? Why indeed, do they arrive like guests, unannounced and unexpected? You see, the gods arrive in the guise of guests, that is our belief. *Athithi Devo Bhava.* The Upanishads say so, and you have seen your own mother give everything to honour self-inviting kith and kin, even though she may herself not have anything left to eat (and the visitor will never guess that). Mind you, guests are gods, and hospitality is a rite beyond duty, and so we wash the feet of the arriving guest, a gesture as potent as sharing one's meals with the gods themselves. Now and then, some irritable visitor might get angry and leave behind a terrifying curse on you. The host, naturally, strives to avert such hostility. At other times, sated and pleased like the intemperate sage Durvasas himself, the guest may shower prosperity not only on you but also your progeny. Yet I ask, how and why does it rain?[4]

The Americans do scientific research, experiments, calculations, simulations, and all that, and publish a plausible explanation for nearly everything. Not only that, like Faust let loose at the frontier, they seem to perform the most wonderful magic in the heavens. "*Nur Einen Schritt, so bist du frei!* One step, and thou art free at last!" is their promise. They say they can make rain through artifice; some say by lighting incense sticks up in the sky. The burning sticks release smoke, so they say, laden with rich silver iodide particles. Some of those clouds, especially those dark-hued ones, contain super-cooled water that will turn into ice when it finds particles to cling to and freeze upon, and furthermore release heat, causing winds, which in turn, they assure us, will make the clouds shed rain. But some other American meteorologists beg to differ and wonder if burning incense sticks indeed causes rainfall. Perhaps we ought to try chanting and dancing too, they say. And those unerring statisticians are still sceptical.

Gelingen und Verdruss
Mit einander wechseln, wie es kann;
Nur rastlos bethatigt sich der Mann.
And worry and success,
Alternately follow, as best they can;
Restless activity proves the man![5]

However, our elders say that if one, well-qualified and pure, selfless and self-assured, did the right Vedic rites impersonally, claiming no rewards, desiring no awards, uttering the incantations, intoning the sounds with the same precision you might use in writing computer code, and placed appropriate gifts dipped in clarified butter in an august fire, and thus caused enough sacrificial smoke to spiral heavenwards carrying vortices of colloidal oblations, then rain shall fall, even if there had been no sign of clouds all year. Prosperity and peace reign where the smoke of sacrifice rises, and the eternal stream of wealth, *yasordhara*, flows endlessly from the gods to us humans. So it is that all gifts are given with the words, "This is but smoke."[6]

Berkeley, CA, 1994

How, indeed how, to invoke the Rain Clouds of Dharma, blue-dark clouds, ferrying rain, heart-warming rain that drops like compassionate tears? Perhaps you can hear, with your extra-sensitive scientific ears, the voice-in-the-sky booming:

> The king of Dharma I am,
> born to crush becoming,
> Like a great cloud
> filled with water, wreathed in lightning,
> resounds with thunder, and refreshes all creatures.

> The Tathagatha I am,
> best of men, a Jina,
> arisen in the world
> like a rain cloud.[7]

Perhaps you, most venerable scholar, do think that Eastern mysticism is only an unexplained phenomenon that is unnecessarily mystified and so do persuade the National Science Foundation and the Social Science Research Council and the Learned Societies to let you sponsor a massive *Yagna*, Vedic ritual, to invoke the rain gods in front of your TV cameras. And so at the Agnicayana, the Namboodri Brahmins bless the Learned Societies, "May they publish the Truth! May they all be blessed! May they be tenured! May all be tearless!" and so on and so forth. Then they solemnly ignite the ritual

201

fire and urge the ignition to own everything one has, for the purpose of sacrifice is the ignition of cognition, is it not?[8]

"Sacrifice makes the world go round and around," our elders say. "Agni! Lord of Fire, I laud thee! Please own this! This is for you and not mine! Not mine!!"

"Sacrifice is the principle of the life and soul of all the gods and all beings." Saffron-hued flames surge playfully from the vessel like dancing serpents and burn time. Verily, music is made when time is flamed, like camphor. If one still feels the need to exclaim "*Verweile doch, du bist so schon,*" then the television camera can sit dutifully in front of the Veda-chanting Namboodri cantors as an electronic fourth brahmin, observing (and recording) quietly. Almost simultaneously you might well watch the proceedings through the INSAT satellite (Indian designed but Palo Alto-made) sitting in Berkeley, even as the hills blaze beside you in weird wind-fanned flames. The burning bushes and the forest fires proliferate on the hills from Los Angeles to Los Alamos, prompting a procession of divine petitions to the curator of southwestern artefacts seeking repatriation of the Zuni Ahayu'da. Sir! you there by the television, reputed to be so clever, knowing thirteen languages (Indo-European, Athabaskan, and Semitic), what does all this mean?

"Nothing!"	"Zilch!"	"Nothing at all!"
"Phat! Hut!!"	*" So Hum!"*	*"Ha! Ouhaa!! "*[9]

To Delhi, to Delhi, to meet the mighty Moghul: 1962

Nineteen sixty two was the year of the silent spring. The makers of poisonous chemicals tried to discredit Rachel Carson as an unhappy spinster with nothing better to do than worry about non-existent children. Fearing a declining demand at home, and seeking an ever larger global market share, American firms, full of febrile and fertile brains nursed in the Second World War, promoted the use of toxic chemicals – pesticides, fertilizers, and insecticides – in the developing world. Thousand-year-old wheat fields in the plains of Punjab were force-fed with concentrated chemicals and dammed water. To ensure this, high officials of American corporate philanthropies had spent a decade assuring Indian leaders that chemical fertilizers would do no harm. Few had heard of Rachel Carson in India, even as the Indian rulers forgot that the fertility and the prosperity of the country depended on the king's virtue. When justice prevailed, the monsoons arrived. As the wise scholar Ananda Coomaraswamy explains, "When a king's virtue fails, the order of Nature is disturbed."[10]

Nineteen sixty two was also the year that brought India running to the Americans for help in utter panic. It was a fateful year. What caused a proud Nehru to beg the Americans for help? Chinese comrades, fellow Asians, annoyed at Nehru's posturing as a world leader, decided to teach him a lesson. They invaded India. Nehru was surprised. Like an ageing Moghul monarch overthrown by his own brother, he was bewildered and exhausted. He would have much preferred rhyme to rhetoric, especially those haunting verses of Ghalib and Iqbal; instead he had to respond to irksome midnight communist communiqués. An Indian journalist asked him:

"Is it still the Indian stand that our frontiers are not negotiable?"
"That is our stand. At the same time there is nothing that is not negotiable," he replied.[11]

The Indian army had been caught utterly unprepared for the Chinese invasion. Ill-clad, thin-blooded, and armed with antique Enfield rifles, soldiers from the tropical plains suffered on the high Himalayan ridges. Suffering suffered, one might philosophize, as the invasion pricked the bloated ego of an entire civilization. Gandhian non-violence mated with Asian nationalism, ending in Nehruvian disaster. He had to seek out Chester Bowles. Bowles insisted on a written request. On 19 November 1962, Nehru wrote to youthful President Kennedy asking for twelve squadrons of US fighter planes and two squadrons of bombers.[12]

Chester Bowles was then back as ambassador to India for a second term because he could not work any longer under Dean Rusk in Washington. Bowles had been a trustee of the Rockefeller Foundation when Dean Rusk was its president. In a certain sense, the hierarchy had inverted when they both went to work for boy Kennedy. Bowles had hoped to be the secretary of state (instead Dean Rusk had been offered the job, it is said, because Kennedy wanted to run foreign affairs himself) and tried his best to be content with something less. In the end, big Bowles chose to relive his glory days as the American plenipotentiary in India.[13]

October, 1951. New York (via Paris, Geneva, and Bombay) to New Delhi

His stay in India ought to have begun splendidly, befitting the Ambassador of the Free World. But the world, Sage Sankara has said, is like a city seen in a mirror, an alien city seen lit at night in a rearview mirror. At two o'clock in the morning of 20 October 1951, Chester and Steb Bowles saw the lights of Bombay from the TWA turbo-prop. It had been a forty-four-hour long journey. The quick halts in Paris and Geneva for nourishment, physical and spiritual, had made the children only more tired. (Remember, all this happened before jet lag was invented). As Chester Bowles recalled:

We woke the children, ran a comb through their hair, put the kitten in her basket, and stepped out into the hot, humid Asian night, doing our best to appear fresh, untroubled and delighted to be in India.[14]

203

The Bowles had read their post reports on India assiduously. India meant maximum hardship allowance, a bonus of 25%. The post report "gave you the feeling," Bowles recalled, "that you couldn't buy anything there; there was no food to be had that was worth eating or no clothing that you could purchase, nothing for your home, and so on." Therefore the Bowles arrived in the sultry heat of post-monsoon October, loaded with "all kinds of canned goods and powdered milk and all kinds of things and clothes to last for five years." They had left behind their lovely Persian mother cat in Connecticut but instead brought along its pretty kitten. Unlike the kitten which knew to hang loose from the mother cat's mouth, they clung together tightly, like baby monkeys do.[15]

> There were a lot of people from the Consul General's office to greet us and pile us all in cars. We started driving through the back streets of Bombay. We saw for the first time – I had never seen one – a big Asian city with all the people sleeping out on the sidewalks – hot, with the terrific hopeless poverty (the post report speaking yet again), or it appeared to be hopeless in the big Bombay slums that you drive through from the airport. And then we arrived in shocking contrast at this great hotel there, a huge place, and were taken into the Viceroy's suite, and there were about eight or nine great rooms assigned to us. Each of the children had a room. It was a huge place. And we finally, *all for safety's sake*, as soon as all the bearers and servants and officials left, we all kind of huddled in the same room, and *slept on the floor*.

Nobody had expected the Bowles to "go native" so quickly. To know the native one must understand the traveler.[16]

Some travelers are angels. "An angel has two understandings. The one is morning light, the other is the evening light. In the evening light he sees things in the light of his own nature." "All for safety's sake," at the Taj! Indeed, in the dark of the night, when all the cows are black, the rope may turn into a serpent and bite you dead. But at dawn, the very same city looked different. "People were active and looked as though they were going somewhere and doing things," Bowles said. He too was going somewhere: *to Delhi, to Delhi, to meet the mighty Moghul*. . . on a majestic US Air Force plane to meet Prime Minister Nehru. He was anxious but also excited. President Truman had instructed him while bidding farewell, "The first thing you've got to do is find out whether Nehru is a communist." (Nehru spoke of Mao respectfully as "Chairman Mao" – not "Mousie Dung" as Truman did.) Therefore Bowles would not have been surprised to discover that Nehru was indeed a communist, but nothing had prepared him for what happened at their first meeting.

"How do you do?" said Nehru, the prime minister.
"Fine, how are you?" Bowles, the ambassador, replied.
"Fine."
Nothing happened.
The ambassador tried to make small talk.
"Well, it's warm weather for October."
"Yes, it is warm. How is it in the United States?" asked Nehru.
"Not so warm there; the leaves are turning."

Nehru was taciturn. Perhaps he was in an introspective mood. His eyes looked inwards. "(They) sort of went up under his eyelids in a strange sort of way, so you hardly see the pupils," the ambassador remembers. Bowles was disconcerted; it seemed as if the great statesman had gone to sleep at their first meeting.

The Light of Asia was just relaxing; letting the mother cat take charge. But how would the American ambassador be expected to know that? Loy Henderson, his predecessor in Delhi, had not briefed Chester Bowles on the mother cat; he had left Delhi in a hurry for Tehran, as one newspaper put it, to "throw water over troubled oils." Furthermore, Ambassador Henderson of Arkansas had never heard of the mother cat. Perhaps the mother cat is imperceptible in the mid-middle of America? Perhaps. But he could have at least warned Bowles about Nehru's empty eyes. But then Henderson disliked Bowles and detested Nehru. Wonder if "Mr. Foreign Service" liked the Shah of Iran any better than the Badshah of India?[17]

The Ganges of clouds

Bundelkhand, 1966

"We, respected sir, in the sun-parched plains of India, watch the skies anxiously, wait and wait day after day, for the herd of elephants to come flying in. Auspiciously rounded grayish-black bodies, big and beautiful, water-bearing, ambrosia-carrying clouds amble over the horizon, urged by the music of South-West winds. As they play in the celestial Ganges our dry fields are sprayed with cool water, Vedica Ap, nourishing aqua. Sometimes, the heat drives them so mad that they rush in without much warning in the morning and discharge their precious fluids. So we invoke Indra's elephant, the majestic Airaavatam, to herd the water-sporting elephants gently.[18]

"Our gods, say, Indra and Airaavatam, Mitra and Varuna, or Rahu and Ketu, are symbols. I have heard a great orthodox Brahmin, well-read in theoretical physics, say that our gods are symbols of natural forces. They are not, he said, just metaphors, but equations of nature. He also said that scientists and engineers in America work with magical mantras and yantras. You probably know about them, formula-barmula, they call them, I am told.[19]

205

"Here, in Bundelkhand, we learn to watch the skies, and observe the clouds play hide and seek with the sun, all year long, moon after moon, season after season. In the summer, after the harvests, the village scholar recites from the Puraanas. I like the catalogue of clouds in the Matsya puraanam."

"The clouds called jimuta make life. Those clouds remain suspended in the air, they change shape, going up a *yojana* to form rain, hence they are the source of life.

The clouds *puskarabartaka* born from the wings of mountains hold a huge volume of water. The cloud named *iva* helps the growth of beings by enhancing the amount of rainfall. And sir, to the north are the pundra clouds which also greatly increase the stock of rain, and all rain produced there becomes, of course, snow.

The clouds, fed by the sun, assume various forms, make thundering noises, produce deluge of rains and quench the destructive blaze of the summer fires.

The chief source of nourishment, without doubt, says the *puraana*, is smoke, especially the smoke of the sacrificial fires. The rising smoke forms clouds. The sun is the centre of the clouds and he absorbs water by his bright rays, from the oceans and lakes. The wind god rides the swollen cloud like a neighing horse."[20]

Smoke of the cremation grounds

Bihar and Bengal, 1948

Historians of Great Britain like to say that the empire was acquired in a fit of absent-mindedness. However, in August 1947, that empire was forgotten in a nervous breakdown. The British, wrecked by the Great War, scampered for safety six months ahead of schedule. Pakistan, two of them, an East and a West Pakistan, were partitioned from India. The amputation was accompanied by an orgy of violence and forced a million Hindus to flee to secular India from Islamic Pakistan. Muslims from India also fled to the Pure Land.

British India had supported the Allies fully in World War II, and recovery was hard for newly independent India even without the burden of the Burra Sahibs. Veterans, traumatized and mutilated, returned to their emptied villages from the minefields of North Africa, and European trenches, and Burmese jungles. But seed grains in the villages had already been commandeered to feed the Allies, causing an enervating, emasculating famine to spread all over the country. The unmentionable was heard and the unthinkable was done. Cows were slaughtered, then husbands ceased to be husbands and prostituted wives, later still mothers hawked their children in Bihar and Bengal. The land was famished. The battlefield was blissful in comparison, some veterans felt. Modern war was but a ritual gone awry. The smoke of the guns rose sky high. The wind gods herded the swollen clouds away, far away. The monsoons failed, once, twice, and thrice. Famine devoured culture. Even the smoke of the cremation grounds ceased to spiral heavenwards, for few could afford pyre-wood. Jackals feasted, near and far, as Great Britain refused to release financial assets lent during the war by India. Such were the conditions in the newly emergent Republic of India, early in 1948.[21]

In that newborn republic of India, an imaginative and talented young physicist, Homi Bhabha, persuaded the nascent Indian state to fund an ambitious nuclear energy program. He argued that India needed novel sources of cheap energy to feed its famished. The terrible famine must never ever recur. Having seen the shortages during the War, he was convinced that having "no energy is more dangerous than atomic energy." Furthermore, inspired by the magnanimous Gandhi, Mahatma Gandhi, he believed it was possible to harness the atom for peaceful purposes, such as digging gigantic lakes for irrigation, irradiating food grains for preservation, and producing cheap electricity for rapid industrialization. Hiroshima and Nagasaki seemed impossible in the India of Gandhi. After all, wasn't Nehru a contemporary version of Ashoka, sending baby elephants to distant kingdoms on peace missions, issuing long declarations and embossing edicts on the virtues of non-violence? Did not his junior minister Morarji Desai preach the benefits of vegetarianism to foreign dignitaries? Had not the noble Nehru loudly and boldly proclaimed the principles of non-alignment and persuaded Tito and Nasser to join him? Who could ever doubt India's commitment to peace and to the poor. Why not use the tiny atom for progress? Why, look, even the aristocratic American ambassador and his family have adopted our cuisine and dress! "Chester Bowles Zindabad! Zindabad! Zindabad! Chester-bhai zindabad!" They are not at all like the sahibs from Europe. How interesting![22]

Post-War Washington received Homi Bhabha, the physicist, for his ideas resonated, at least rhetorically, with Eisenhower's "Atoms for Peace" plan. Propaganda was a much used weapon in the cold war. But, given the paranoia of Cold War politics, few accepted India's repeated affirmation of its commitment to peaceful uses of atomic energy. And then the infractions over Kashmir and Goa convinced cynics of India's "duplicity." Bhabha's obvious determination to master nuclear technology ensured a close monitoring of the Indian nuclear program by the US. Idealism is not attributed easily to the poor.[23]

Homi Bhabha

Cambridge, Bangalore, Bombay, 1930s and 1940s

Homi Bhabha created new institutions for the assimilation and adaptation of advanced technologies and, as it were, institutionalized his approach to high technology in post-colonial India. He was so successful that the vision of Bhabha has continued to guide the Indian space and nuclear programs for nearly four decades.

But most Indians have not heard of Bhabha. We Indians don't celebrate heroes; we only worship them. Thus for many younger Indian scientists, Homi Bhabha remains only a remote paternal figure represented by indifferently sculpted, dusty, solemn-faced marble or patinated bronze busts found in the dim halls of scientific institutions. Yet to know and appreciate the life and work of Bhabha is to understand the high technology enterprises of the post-colonial Indian state.[24]

Endowed with an aristocratic bearing, Bhabha was a brilliant theoretical physicist. He was also a passionate collector of art, a star oarsman and vigorous tennis player, an accomplished artist and musician, a dynamic leader of the scientific community, a brilliant administrator, and above all an ardent patriot. Homi Jehangir Bhabha was verily a versatile genius, and a worthy founding father.[25]

Born to an influential Bombay Parsi family in 1909, the young Homi Bhabha attended a private school established primarily for European children. His father, an attorney, expected him to join either the bar or the rapidly growing industrial empire of the Tatas (who were close relatives). In preparation, Bhabha was sent to England, as was then customary among wealthy Anglophile Indian families. He began undergraduate studies at Gonville and Caius College, Cambridge, which counted his uncle Sir Dorab Tata among its alumni and benefactors. There, Homi Bhabha was tutored by Paul Dirac and apparently "took to quantum electrodynamics . . . like a fish to water." Convinced of his own taste and talent for physics and mathematics, Bhabha persuaded his pragmatic father to let him pursue his interests. The elder Bhabha finally relented, and Homi Bhabha began research at the Cavendish Laboratory. Soon after, in 1933, Bhabha published his first scientific paper, at the age of twenty-four, in *Zeitschrift fur Physik*, allaying his father's anxieties somewhat. After obtaining his PhD from Cambridge in 1935, he continued his theoretical physics in Europe, visiting various centres of advanced physics on the continent. He quickly earned a reputation as a brilliant theoretical physicist, and was soon elected a Fellow of the Royal Society in 1939.[26]

Homi Bhabha was in India on a short holiday when the Second World War erupted. Unable to return to Europe, he accepted a specially created readership in cosmic ray physics at the Indian Institute of Science, Bangalore, founded by his relatives, the Tatas. Professor C. V. Raman, who by then had acquired a well-deserved reputation for being an ingenious experimentalist and a brilliant teacher, was India's only Nobel laureate in the sciences, and naturally, he headed the physics department at the institute. With Bhabha's arrival in 1939, the institute hosted the best experimental and theoretical physicists in India. Two years later, Bhabha was promoted to the Professorship of Cosmic Rays and continued his theoretical studies, attracting a small group of researchers around him.[27]

Bhabha's experience at the Indian Institute of Science during the war years was crucial in many ways. C. V. Raman played an important role in convincing him to stay in India. Ironically enough, Raman might have also convinced Bhabha to get out of the cramped world of Indian universities. He decided to concentrate on the difficult task of creating new institutions embodying a new culture of science befitting modern India. Drawing on his own experiences in the elite physics community of Europe, Bhabha set himself the task of creating a pleasant environment for advanced research. He wanted an inspiring environment: one that avoided the petty jealousies and Brahmanical inefficiencies of Indian universities which mired even the Nobel laureate C. V. Raman like an elephant in a tropical swamp. The Tata Trusts, controlled by his close relatives, aided Bhabha in this ambitious task. In 1945, a generous and timely grant enabled him to create the Tata Institute of Fundamental Research at Bombay, "an embryo from which to build . . . a school of physics comparable to the best anywhere." There, amidst landscaped gardens and evocative modernist paintings, Bhabha envisioned the future India.[28]

Being a member of the Bombay Parsi elite by birth, Bhabha came into frequent contact with the leaders of the nationalist movement. The independence movement, ascetic and austere as it might have been under Gandhi's leadership, still needed financial contributions from the merchant-princes of the Parsi community, and so nationalist leaders like Nehru and Gandhi were often their houseguests. These informal contacts with the statesmen of India provided by his family proved important in helping the young Homi Bhabha direct India's high technology ventures.[29]

Jawaharlal Nehru, on his part, carefully cultivated an alliance with the wise men of science to turn India into "a paradise on earth." Of the many eminent scientists engaged in harnessing the powers of science, including Meghnad Saha, C. V. Raman, and K. S. Krishnan, Nehru was drawn most to the patrician Bhabha. By the early 1950s, when India optimistically prepared for rapid "development," Bhabha was meeting "Jawahar *bhai*" Nehru informally every fortnight – a privilege few others enjoyed. The ramifications of such access to the most powerful leader of a newly independent country cannot be overemphasized. C. V. Raman's failure to cultivate the friendship of Nehru contributed to his frustrations in building new scientific institutions. Meghnad Saha's social and political differences with Nehru propelled Saha's entry into mundane electoral campaigns, mapping a sad decline from high physics to petty politics. Bhabha, in contrast, glided easily from elite scientific circles to policy-making at the highest level.[30]

Soon after independence, Bhabha quickly obtained the formal approval to create the Atomic Energy Commission, and under its umbrella, he organized a vast empire of research. He was convinced that even a backward country such as India could catch up with the West in an emerging field like atomic energy, precisely because of the field's nascent character. He often said "Not all aspects of . . . research are so expensive and Indian scientists can well participate with the scientists of other countries . . . on the frontiers of knowledge."[31]

Clouds of war and peace

Lop Nur, 1964

On 17 January 1955, the Soviet government announced that it would help the communist nations, including China, "promote research into the peaceful uses of atomic energy." The Soviets were compelled to respond. The cold war was a subtle war fought on the front pages of newspapers. China cashed in. On the fourth of July 1955, the Politburo appointed a three-member committee to formulate and implement China's nuclear program. Nie Rongzhen, a trusted soldier of the revolution, was chosen to head the program. A special Bureau of Architectural Technology (*Jianzhu jishu ju*) was created to camouflage the atomic bomb program. By the summer of 1958, under the Great Leap Forward, tens of thousands of peasants armed with scintillation and Geiger counters were prospecting for uranium. "The whole people should engage in Uranium mining" (*quanmin ban youkuang*), the slogans of the Second Ministry of Machine Building (in charge of the nuclear program) declared. Teams of scientists and engineers taught peasants in Hunan province to make uranium yellow cake from mined ores. The peasants produced the first 150 tons of uranium concentrates even as the high-technology factories commissioned for uranium production ran into difficulties. The peasants had won the race to make uranium, thus making the first Chinese bomb verily a people's bomb.[32]

Early in 1964, the American government had become acutely aware of communist China's nuclear capabilities. Indians were informed of American reconnaissance flights out of Thailand to monitor Chinese activities in Xinjiang province. The Central and Defense Intelligence Agencies were particularly interested in developments in Yanqi.[33]

Fifteen hundred years ago the Buddhist monk Faxian had met several monks from India among the thousand monks of the Hinayana school here, at the ancient town of Yanqi in the Gobi. (Later still Faxian himself traveled to Gaya and Nalanda to study Buddhism.) Many centuries later, in 1964, American planes saw increased human activity around Yanqi on the western edge of what is now the Lop Nur Weapons Test Base. The secretary of state, Dean Rusk, publicly announced the forthcoming Chinese atom bomb causing "great disquiet" in India. The Indian ambassador in Washington, B. K. Nehru, met senior officials of the State Department ten days before the impending Chinese nuclear explosion. Indian security was at stake. The silk route, which had seen caravans carrying corals and lapis and myrrh and frankincense from Benares to Beijing, along with Buddhism and metaphysics, was fast turning into a busy military highway through the Karakoram mountains, allying China with Pakistan. (The Chinese conquest of Tibet had ensured that.) Moreover, the invasion of 1962 was still fresh in public memory. The Peoples Liberation Army could have gone all the way to Gaya like Faxian, if it had wanted to do so. But they were no pilgrims to the holy land.[34]

But of all things, Ambassador B. K. Nehru (a cousin of Prime Minister Nehru) pleaded for a public certificate from the American secretary of state, stating that

> India like Communist China has the potential to produce nuclear weapons but as a good citizen of the world, India has no intention of producing nuclear weapons; (and that the) US commends (the) path chosen by India which contrasts with (the) course being followed by Chinese communists.

B. K. Nehru urged the Americans to issue the statement while the Cairo Conference of Non-Aligned nations was still in session. He felt that "such a statement . . . would have considerable effect on Afro-Asian governments." (Nehru's India had been vying with China for the leadership of the Third World. Chou Enlai, in turn, had felt insulted by the Indians at past Asian conferences.) Ambassador Nehru said that such a certificate from America would "redress the psychological balance to considerable extent." Strangely enough, some high officials of India perceived the Chinese atom bomb to be a public relations disaster rather than a palpable security threat to the country. Who in Delhi knew the war widows in Madurai?[35]

In October 1964, two years after the invasion of India, China successfully exploded a twenty-kiloton nuclear weapon, causing consternation south of the Himalayas, in Bombay. India's sole friend Khrushchev had just died. India also found it was not as popular as it had imagined at the Cairo Conference of Non-Aligned Nations. The Uranium Sutra was read in Bombay, even if Delhi insisted on talking about the Lotus and Diamond Sutras. India's Nehruvian Sinophilia vaporized completely at the Department of Atomic Energy. Homi Bhabha was determined to ready an atomic weapon in eighteen months. He was no Nehru; held no romantic notions about China.[36]

Four months later, the American embassy in Delhi sought to "mitigate the impact" of an impending second Chinese communist nuclear explosion on India. The embassy suggested that the prime minister of India

> be allowed if he wishes to do so to announce the impending CHICOM (Chinese communists in cold war lingo) blast, or failing this the blast itself. The purpose of this would be to suggest that India can monitor such CHICOM nuclear activities, thus enhancing India's apparent scientific capability.

The telegram added that India's new prime minister, Shastri,

> is personally determined not to undertake a nuclear program for India, but he badly needs reassurance both at home and abroad. . . . We still have a rare opportunity to help Shastri keep India out of the nuclear weapons race if we act promptly.[37]

The humiliating defeat in the 1962 China war and China's nuclear explosion of 1964 severely tested India's faith in Gandhian non-violence. Indians, Raja Rao wrote:

> live history unhistorically, and play revolution with non-violent violence or play Mao with and against Gandhi. Showing the power of the spinning wheel against the atomic reactor we're poorly clad against China's might. Friends of all, and having no friends, we spin our way through time on speeches. Unable to use force against force, we play chess at the U.N. We have forgotten that chess is Indian in origin, traveling these twenty centuries through as many countries, it has changed its rules. So we play neither according to their rules nor ours.

At least Bhabha understood "their rules." The 1965 Pakistan war further strengthened his position in favor of an Indian nuclear weapon. Earlier, Nehru's categorical rejection of the atomic bomb had stood in his way. Nehru's death, although a significant personal loss for Bhabha, brought Lal Bahadur Shastri to power. Shastri was well informed about atomic physics. He had even translated a biography of Ève Curie into Hindi. He permitted Bhabha to hasten weapons research. Bhabha hurried towards an Indian nuclear bomb. Ambassador Bowles promptly reported to the American leadership,

> the Indians as you know could detonate a nuclear explosion within a year or so. . . . This is partly for prestige purposes and partly because of a genuine fear that the Chinese or Pakistanis would be just crazy enough to blow up Calcutta.

Late in November 1965, Bhabha publicly declared that India could produce a weapon within eighteen months from the word "Go!"[38]

How to keep India out of the nuclear weapons race? How to goad India into using its scarce resources for development? How to soothe India's anxieties about its status in the world? Such questions kept Chester Bowles awake at night typing lengthy memoranda to Washington. Noble and compassionate, he was concerned about feeding India as much as he was about securing American interests in the region. It is true, he did worry about India more than Washington was wont to do. He was an American idealist caught in a "cloud war."[39]

The drought of 1965 was crippling; one of the very worst in a century. Perhaps, Indian farmers could start using more fertilizers. Perhaps, the Rockefeller Foundation could promote a green revolution to keep the power-hungry communists away. Perhaps, American space technology could be used to help the Indian farmer. He was a dreamer, a visionary, and so a misfit. "India needs a great technological success for all the world to see. Could we . . . dig a vast crater to control the flood waters of the Jammu?" he wondered.[40]

The war with Pakistan in 1965 had further hurt the Indian economy, although India had acquitted itself well in the war. (As one Indian general observed, it was more of a "communal riot with armor" than a war.) The humiliation of 1962 gave way to jubilation, and Prime Minister Shastri pursued peace with the confidence of a victor. He flew over the Arabian Sea, Iran, Afghanistan, and around Pakistan to reach Tashkent to make peace with the Pakistani leader, General Ayub Khan. Shastri negotiated from a position of strength. But he died right there in Tashkent, it is said, from a heart attack, on 15 January 1966.[41]

A few days later, Homi Bhabha had to attend a meeting of the International Atomic Energy Agency in Vienna. Shortly before he was to leave, Canadian nuclear energy officials called. They sought to meet with Bhabha in Geneva before he proceeded to negotiate with French nuclear experts on the transfer of technology. The Canadians, it seemed, were willing to outbid the French. Bhabha changed his flight plans at short notice. He boarded an Air India plane on 24 January 1966. A mountain rose into the skies to halt Hanuman as he flew to Lanka, so says the Ramayana. Homi Bhabha was no Hanuman, though. Air India "Kanchenjunga," piloted by one of the most experienced pilots, flew right into Mt. Blanc, failing to clear the summit by less than fifty feet. The only survivors were a few miserable monkeys, brothers of Hanuman, on their way to American laboratories for scientific research. News of the accident reached Indira Gandhi shortly after she was sworn in as the prime minister. Within a fortnight, India lost two of its gifted leaders:[42] "*yat purushena havisha.* . . . The creator is the sacrificial victim. . . ."[43]

Ambassador Bowles had an intimate view of the shock waves that the two deaths in two weeks caused in India. He had hoped that Chavan would succeed Shastri. Chavan was known to be particularly friendly to America. But he was not particularly liked by the Congress Party, and India was a democracy. Either Morarji Desai or Indira Gandhi was likely to be elected the new prime minister. Morarji Desai, a strict moralist, had made many enemies. For instance, he had completely alienated Dean Rusk in 1957. "It looks like Indira Gandhi," wrote the American ambassador, adding perceptively, "Indira frightens me a bit. But she might rise to the task. Much will depend on whom she depends on for advice." He, the representative of Uncle Sam, volunteered. He remembered that she had responded well to his avuncular advice earlier. It was rumored in the bazaars that she was getting large amounts of money from the Soviets to win the elections. Indira Gandhi was not eager to be tagged "Pro-Soviet." She let Bowles help.[44]

To his credit, Ambassador Bowles, it must be said, did succeed in persuading America to assist India. One of his suggestions involved making artificial rain. "Rain," Bowles noted in his Delhi dairy, "is food, security, hope, promise. In our industrialized world it is . . . postponed golf games and general inconvenience. In India, like everyone else, I love the rain and find it wonderfully exciting and reassuring." If it worked, India, he thought, would be grateful to capitalist America for soothing the vagaries of the monsoon. If the rain-making technology worked, it would also boost India's confidence in the power of Western science and technology. The idea was received warmly in certain sections of the American establishment, particularly the Department of Defense. The American armed forces had been toying with the idea of cloud-seeding to induce heavy rains in Vietnam to impede the movements of the Vietcong. A special secret experiment named Project POP-EYE was planned along the Ho Chi Minh trail. The new monsoon cloud-seeding technology needed a laboratory. India was a suitable site. Project POPEYE was quickly re-christened Project GROMET and taken to India. In case news of the experiment leaked, the change of name ensured it would sound vaguely green-revolutionary.[45]

Project GROMET was a classified experiment conducted by the US armed forces in India early in 1967. The Gangetic heartland of India had seen several monsoons fail and was caught in a severe drought. Although the climatological conditions for artificial rain-making were not "very favorable in Bihar-Uttar Pradesh during January but much more favorable during the period May-October . . . we (Department of Defense) are prepared to consider program in Bihar-Uttar Pradesh area beginning late January in an attempt to improve crop yields," Robert McNamara's deputy wrote Chester Bowles in late December 1966.[46]

"Improving crop yields" was a constant mantra heard in 1966, because India had approached the United States for a loan of food grains. Indira Gandhi had been summoned to Washington in March to plead her country's case. She did her best to swallow her pride and traveled to America to meet President Johnson. LBJ toasted Indira Gandhi, saying: "Missus Gaaandhi, Ah only hopes we can somehow be worthy of ya." But LBJ being LBJ, it meant more than what it said. Was Indira Gandhi going to be worthy of him? Will she support him on Southeast Asia? "Hell, no!" He was not going to send food grains to those who refused to fight with him on Vietnam. "Like a little boy pulling the wings off a butterfly," he cracked India's pride and made Indira Gandhi's government succumb to American demands. She had little room to maneuver between the Americans and the Soviets. Will she hang on? How? She is young and frail, constantly falling ill. May the gods help her![47]

Delhi consented to Project GROMET. It moved very quickly through Washington. LBJ was personally interested. On 29 December 1966, Walt Rostow reported to the president that

> a DoD meteorologist, just returned from New Delhi was impressed by his Indian counterparts. The planning and execution will be a

thoroughly joint enterprise. Three or four planes will be used with commercial markings, flown by Americans in civilian clothes. Indian Airforce meteorologists will participate in each plane.

He also informed Lyndon B Johnson that "an agreed contingency press release will characterize it as an agro-meteorological survey, in case of need." LBJ was desperately in need of new technologies to win the war in Southeast Asia. As early as 1957, he had considered military applications of weather modification. Someone had told the senator from Texas that the total amount of energy released by a single thunderstorm could equal that of a twenty-kiloton atomic bomb. Power – raw and ruthless – fascinated the Texan.[48]

By 1 February 1967, American military planes, C-130s, were ready at Indian airfields, having flown their first local test flights and waiting to start their experiments. The National Security Council reported to Walter Rostow at the White House, "When and if there are any results, I will let you know right away." Three weeks later, after several attempts, Rostow sent a note to LBJ:

> As you know, we are making a rather desperate effort to make a little rain fall in India at a time when there aren't many clouds about. . . . We cannot really expect much until the monsoon clouds appear in mid-year.

By the end of the month, the US military had flown several missions over India looking for clouds to seed. The heavenly elephants had gone sailing away (and the goddess of prosperity is seated on them). On 28 February 1967, Rostow concluded for the president,

> despite the limited results of this experiment, . . . economically useful amounts of rain can be produced over much of India during and after the monsoon season when there is usually abundant non-raining cloud cover. State (Department) and the scientists are sorting out what kind of a statement to issue – if any.

The results of the hundred rain-making experiments were, at best, ambivalent. Senator Pell concluded that "an elephant labored and a mouse came forth." Yet the White House sought to turn them into a cold war triumph, making hay as the sun shone.

> We and the Indians have carefully worked out agreed answers to possible press queries, . . . there are good reasons for postponing any comment for as long as possible. When a statement is to be issued, it should be by LBJ, as it is a very important breakthrough. We may even want to hold it until the Water for Peace conference, when it might be a central U.S. gimmick.[49]

Unfortunately artificial rain-making remained just a gimmick. When the monsoon season arrived in July, and the celestial elephants returned, artificial rain-making met with no greater success. Without a trace of irony, a White House official reported,

> Rainmakers bog down. . . . As you know, from the beginning the scientists have been skeptical of the results obtained. They felt it necessary to insure "scientific" control of the results. . . . The Indian scientists with whom they worked were even more purist on methodology. . . . This appears to be the cost of moving it from operators to scientists. The latter could not be infused with a sense of responsibility to promote Indian rains if there might be some cost to scientific purity. The Embassy worked hard on the Indians but could not persuade them that our approach was sufficiently promising.

The drought-devils remained ensconced in their sandcastles. Although one may catch an elephant through guise and guile, seeding them is entirely another matter (as American zookeepers well know). Anyway, the celestial elephants declined to be impregnated by the American C-130s in disguise.[50]

Cloud of unknowing

Bundelkhand, 1966

Highly classified experiments in military weather modification to ease the drought in India were all for safety's sake. Whose safety though? So if you, the director-general of the Department of Meteorology in New Delhi think that the Americans are not being scientific enough, and you have "put papers" in *Science* and *Nature* to back your claims, you get transferred to a waterless scrub jungle. You meet Him, and He, your long-waiting teacher, introduces you to a village scholar. The village pandit is getting old, we all are, and he seems to be forgetting some of the *puranaas*; his son has left for the city, leaving us to wonder if those German-sounding machines purchased with NSF grants will remember the sacred lore. A French brahmin came to record some more of what the pandit knows. He even left a copy.

> "Yea! you there, yea! ye! Corner-house-Kanha, go ask Shankar's-mother to give you the tape and the machine. Make sure you don't drop it."
> He played a ringing Rg Vedic hymn addressed to the clouds:
> *accha vada tavasam . . .*
> *pra vata vaanti . . .*
> *divo no vrshtim. . . .*
> *kumuta prajabhyovido Maneeshaam.*

The machine hymned just as a pandit might –
I address the mighty cloud,
with these invocations; etc. etc.[51]
Maybe there is something to all these Vedic utterances, you begin to
wonder: your scientific curiosity is piqued.[52]
He, on the other hand, shows you some other experiments. "Ask me
questions. Challenge me if I am wrong," He says, "learn to unlearn."

More pots and pans

New Delhi, 1950s

On his first tour of duty in India, Chester Bowles established a tradition
of distinguished American ambassadors concerning themselves with India's
economic development. In the early 1950s, Bowles met with Nehru two to
three times a week, week after week, usually discussing a broad range of
subjects ranging from international politics to Buddhist history. In one such
meeting early in 1953, ironically enough, the American ambassador ended
by defending Stalin: "No man could be that bad." "Well, this man is," said
Nehru. Mostly Bowles stirred a shared enthusiasm for talking about modernization. "Well, he (Nehru) used to talk development with me, you see,"
recalled Bowles.

> I was all involved in the whole economic development of India and
> the social development and the political development. He almost
> took me as a partner; this was my relation to Nehru, basically this
> commitment to the development of India, and he thought I knew
> enough about economics to be helpful . . . almost every embassy
> in Delhi used to come to us to find out what Nehru was thinking.

Nehru was wont to pick the ambassador's brains on issues such as rural and
industrial development. He would want to know, for example, "whether
the biggest answers came from the giant dams or whether they came from
small waterworks and tube wells." Bowles used to discuss all these things
endlessly. He was concerned with "this great question of how you involve
the people and bring them in as participants in the whole approach to
growth."[53]

Sustained talk of development contained communism in the Third World.
Walt Rostow had theorized about the stages of development. But British-
educated Indian leadership was in a hurry.

Vikram Sarabhai, Homi Bhabha's illustrious successor, has said "there is
no hope for those who wish to advance step by step following outmoded
methods." Maybe we Indians should leapfrog. The Maharishi might teach
us all meditation and then, like Hanuman, we too can leap across the

oceans, jumping high up from a bullock cart on to an over-flying Boeing, if not to a heaven-reaching elephant. Why, anything was possible with our noble Nehru in the cockpit with his grandsons. In the early 1950s, Anglophone India, having recently inherited the Raj, was optimistic and enthused about India's future. Anglophone India found the Americans to be very different from the British. Some of the American experts who came knocking at the durbar doors, it noticed with some concern, could infect India with their democratic enthusiasm. Chester Bowles was certainly one of the earliest of those American experts. In New Delhi, the doors of the prime minister's house were remarkably open to American experts, if not American expertise.[54]

Strange as it might have seemed to the courtiers in Delhi, considerable American know-how was housed in their profit-making, dividend-declaring, stock-splitting corporations. The New Delhi durbar was understandably proud to entertain White courtiers but not corporate cowboys. The cowboys were often prairie people, bringing the frontier with them, barbed wire and all, veritable Vasco da Gamas of the grasslands. Even the American State Department was embarrassed sometimes by the cowboys. To make matters worse, the frontiersmen liked their Indians rarely and, in turn, Nehru's Ox-Bridge Indians reciprocated heartily. So the Malayalees in the Ministry of External Affairs handled the Vasco da Gamas from the New World.[55]

Nevertheless, certain basic paradigms in the practice of American foreign policy in India emerged from the perception of India as an open laboratory for development. Indians, Chester Bowles found out early on, did respond well to American idealism. Some ambassadors, like Loy Henderson, and the foreign policy establishment in Washington did not share that same liberal enthusiasm for Indian economic development. John Foster Dulles, for instance, told three hundred Americans in Delhi on his first trip in 1953, "leave them alone for a while and the Indians will come crawling to us on their bellies." Washington, on the whole, tended to be pragmatic in its thinking, and *realpolitik* dominated its practice.[56]

In the end, India was interesting only to the extent that it could prove itself a counterpoint to communist China. "We used to talk about China where everything was done autocratically, where people were more or less pushed into these things, and their enthusiasm was whetted by all kinds of slogans," Chester Bowles recalled.

> And what was democracy's answer to that, and I used to argue that to a degree it had to be consumer goods, and this always meant that any democratic country would have a slower growth rate, because you couldn't get the capital investment and still get the more shoes, the more bicycles, the more pots and pans that you needed to motivate people.

Indian economic development thus became a genuine albeit minor concern of American foreign policy early in the cold war. The "loss of China" had made India important, sort of, to the cold warriors.[57]

Preventing mushroom clouds

Bundelkhand, 1990s

How to get more shoes, more bicycles, more pots and pans? How does one keep the pots brimming? In what manner is abundance attained? Things, Proust says, are gods. Why? How, Oh! Honoured one, to obtain the treasures of the earth?

> May Earth with polyglot people,
> of religions varying
> according to places of abode,
> pour for me a thousand streamed treasure
> like a never failing cow.

So sang the Seers. What was their method? If their Truth is eternal and beyond history, is their path mapped? Have you read Raja Rao?[58]

> What Gandhi did in politics, and Tagore in poetry, Ramanujan performed in mathematics. But he died before he could tell us more of his method. And that is what our job is – the job of the Indian scientist today – rediscover the Indian methodology of research.[59]

What made India splendid and fantasy-provoking? What drew Alexander from Macedonia? What brought Chester Bowles to India? Was he a development dean, a secular successor to the machine-operating Bible-quoting missionaries from America stationed at Allahabad or Etawah? When and why did India get poor? What is the history of Indian indigence? There is always enough time for such speculation in Bundelkhand.[60]

India's poverty is recent. We were mostly a prosperous land, no matter what your imperial historians tell you.

Where did India's wealth come from?

From within. We never sent armies to conquer and loot.

How did India become rich and famous through history?

By being India. By being true to herself.

What do you mean?

You see, America became wealthy by being faithful to its principles, by following its own dharma. Similarly, India's wealth will follow our search for the Truth. In the Mandukya Upanishad is said *satyameva jayate*, Truth alone triumphs, and that is our motto, is it not?

How does one search for Truth – a breath, a wind, a shadow, a phantom?

By going to the very end of all your questions till there are no more questions left. Remember Nachiketas went all the way to the nether-world and waited three days for the lord of death, Yama himself, (green hued and clothed in red, long noose in one hand, terrible maze in the other, riding from the south on a water buffalo). You may certainly go to the forests of Penn. I shall talk to your grandmother, if need be; she will bless you. Please remember that Nachiketas turned down Yama's offer of "fair maidens seated on chariots and playing celestial music, the like of whom men verily cannot obtain," to seek his answers. Like him you too may have to learn to say, "let thy vehicles, thy dancers and singers remain with thee!" Never, never ever stop till all your questions are answered.

Having known that, whatever one desires, one obtains.

Yes-yes. That is indeed what the Upanishads say. But look, in our own time, Gandhiji achieved great success by following the Indian way. His simple satyagraha shook the viceroys of Imperial Britain. Satyagraha is the grasping, the assertion of Truth. And remember Gandhiji saw nothing negative. Recall his three monkeys. He urged Indians to love the Englishman, not to hate the British. He was a universal man, a mahatma. It is said he had the power to press heroes out of clay.[61]

Why did Gandhiji anoint Nehru his successor? They hardly concurred on substantive issues. "Gandhi was really a Hindu. Not at all a gentleman, like Nehru," Andre Malraux noted. Gandhi identified himself totally with India and the Indian peasant. On the other hand, it was easy even for an outsider less astute than Malraux to see that there was "an enigmatic distance between Nehru and India." Interestingly, the two held significantly different views on industrialization and mechanization. Yet Gandhi said of Nehru, "I know that when I am gone he will speak my language." Gandhi, apparently Luddite, saw the limits and consequences of global industrialization. Nehru, mostly reacting to colonialism and Western imperialism, saw science and technology as a way for India to catch up with the West.[62]

"It is machinery that has impoverished India," Gandhiji declared. "It is difficult to measure the harm Manchester has done to us. It is due to Manchester that Indian handicraft has all but disappeared," he wrote. As a remedy, albeit symbolic, he urged his followers to spin and weave. He even cajoled Maurice Frydman, an inventive Polish electrical engineer working for the Maharajah of Mysore, to improvise the archaic village spinning wheel. Gandhiji talked of both *swadeshi* (of-one's country) and *swaraj* (self-rule) for he knew, like most Hindus, that prosperity was married to *swaraj*. But how can the archaic be modernized?[63]

Nehru, forgetting the innate pragmatism of Truth, grew anxious about Gandhi's technology policy. "The spinning wheel is not as strong as the machine," he bemoaned to Andre Malraux. Maybe had Nehru been spinning religiously in Delhi, Chou Enlai might not have felt slighted at Bandung, and the need for French fighters and American bombers may not have arisen.[64]

"Machinery has begun to desolate Europe. Ruination is now knocking at the English gates," Gandhi wrote several years before World War II. "Machinery is the chief symbol of modern civilization. . . . If the machinery craze grows in our country, it will become an unhappy land," he said. Nehru, of course, did not agree with Gandhi's attack on machinery or modernity. Gandhiji later softened his critique of technology:

> even this body is a most delicate piece of machinery. The spinning wheel is a machine; a little tooth pick is a machine. What I object to is the craze for machinery, not machinery as such. The craze is for what they call labor-saving machinery. Men go on 'saving labour' till thousands are without work and thrown on the open streets to die of starvation. I want to save time and labour, not for a fraction of mankind but *for all*. . . . For instance, I would make intelligent exceptions. Take the case of the Singer's Sewing Machine. It is one of the few useful things ever invented, and there is a romance about the device itself. . . . The machine should not be allowed to cripple the limbs of man. . . . Ideally, I would rule out all machinery, even as I would reject this very body, which is not helpful to salvation, and seek the absolute liberation of the soul. From that point of view, I would reject all machinery, but machines will remain because, like the body, they are inevitable. The body itself, as I told you, is the purest piece of mechanism; but if it is a hindrance to the highest flights of the soul, it has to be rejected.

As with other problems, Gandhiji had thought through the question of technology carefully.[65]

After Gandhiji's untimely death in 1948, Nehru could play at large-scale industrialization more freely. His science advisor, Bhabha, had concluded that "what the developed countries have and the underdeveloped lack is modern science and an economy based on modern technology." Therefore "the comfortable bourgeois prosperity that is to be . . . established when we have learned enough science and forgotten enough art to successfully compete with Europe in a commercial war" became his priority. Gigantic dams, petrochemical-based fertilizer factories, state-owned steel plants, and many other such gargantuan ventures were nurtured by Nehru, even as the state deified Gandhism and the idea of Gandhian small-scale, decentralized, village-based economies. "It was science alone that could solve these problems of hunger and poverty, of insanitation and illiteracy, of superstition and deadening custom and tradition, of vast resources running to waste, of a rich country inhabited by starving people," he intoned. In this manner, each year, he, the deacon, the *purohita* of Indian science, quite religiously cheered and led the Indian Science Congress. "I too have worshipped at the

221

shrine of science and counted myself as one of its votaries," he confessed, adding that "the scientific method alone offers hope to mankind and an ending of the agony of the world." The scientist was, like a Hindu *sanyasin*, a "sage unattached to life and the fruits of action, ever seeking truth wheresoever this quest might lead him. To tie himself to a fixed anchorage, from which there is no moving, is to give up that search." In contrast to Gandhiji, he had very high expectations – technocratic expectations – of science and scientists. Why then did Gandhiji anoint Nehru knowing fully well that significant differences separated them?[66]

For many centuries Indian society had withdrawn into itself like a traumatized turtle. Some say it was because India was a besieged land. Whatever may have been the origins of that isolationism, the fact remained that in the mid-twentieth century, India knew little of the world outside, and what little the world knew of India was mediated by missionaries and conquerors. Gandhiji thought young Nehru was the one most likely to think globally and integrate India with the rest of the world. Jawaharlal Nehru was widely traveled, so at ease in the West, and he also took a deep interest in the fate of coloured peoples in Africa and Asia. No other figure in the Indian freedom movement knew the world as well. Sardar Patel certainly understood himself better. But what about the world? The new Indian nation-state had to lead the world. So Nehru became a world leader, a statesman for the colonized and the oppressed all over the world. He traveled far and wide delivering eloquent speeches. "The Emperor speechified."[67]

He, indeed, loved to give speeches. As the great linguist Whorf has pointed out, "Speech is the best show man puts on," and in thinking thus Whorf was not far from traditional India which believes in the essentially vocal character of government. *Vaca kritam karma kritam*, says the Mahanarayana Upanishad. What is done vocally is done indeed. *Yad vava vaca karoti tad etad evasya krtam bhavati*, declares the Jaiminiya Upanishad Brahmana. What is said verbally is what is considered done in deed. Speechifying assumed a renewed importance in post-traditional India, and thanks to the British, the Indians could wield an international language with fluency and eloquence. Vox imperium husbanded vox populi.[68]

Why then were Nehru's irksome speeches and neutralist diplomacy tolerated? Curiously enough, he might have been the only Third World leader of any stature not subject to assassination attempts. "If Nehru goes, Communism takes over," Bowles reported. Early in 1948, Washington consulted Sir Olaf "O.K." Caroe, a governor in British India, on the impact of Nehru's statesmanship on the Muslims of Afghanistan and the Middle East. O.K. was dismissive: "In world affairs, Nehru is an adolescent." American diplomats continued to carefully assess the response of African and Asian peoples to Nehru after every major event. Was Nehru likely to upset the apple cart by inciting nationalist revolutions all over the world? This was a question American diplomats pondered carefully in the late 1940s and early 1950s.

They concluded Nehru was the least troublesome leader that India might have and that his speechifying posed no real threats. Like Nehru, Ambassador Bowles enjoyed giving speeches too and so got along well in India. Of speech, we say, like Rilke, "*du bist gross.*" A big deal, shall we say.[69]

He spoke of American democracy and Indian achievements, of this and that, here and there, now and then. He debated with college students, journalists, and villagers. Unlike Nehru, he was an authentic worshipper of the word; his substantial fortune arose, not surprisingly, from the adoration of the spoken word. "The vocable is not only a big deal but it is gold." Prior to entering politics and diplomacy, Chester Bowles had found both fame and fortune on Madison Avenue selling the spoken word on the airwaves. His experience in wartime administration in Washington had made him a firm believer in the power of technology in selling ideas. "If there's one thing I learned in the advertising business," he told a New Delhi embassy officer, "it was to make people think they have a choice but not give them one." Utilizing his experience in advertising, politics, and administration, Bowles effectively sold American deals, ideals, and images to India during his first term as an ambassador. He launched a highly successful news magazine, *The American Reporter*, in English and fourteen other Indian languages. *The American Reporter* reached remote villages even before Coca-Cola got there. Bowles, a formidable marketer of dreams, dedicated himself to economic and social development.[70]

When Indira Gandhi became a junior minister of information and broadcasting in the Shastri cabinet, Bowles found a sympathetic ear. He, of uninhibited Yankee ingenuity and boundless energy, encouraged her to think of a massive broadcasting network to spread development messages to the villagers of India. Gandhi found the ability to directly reach millions of Indian voters appealing. She appealed to the United Nations to fund her radio project. She sought the help of the master of words, Vacaspati, Chester Bowles to persuade USAID to "undertake a feasibility study to determine the necessary number of transmitters, their cost, . . . and other specifications of the broadcasting system."[71]

In a letter to an American friend of the Nehru family, she wrote in 1965 asking about the possibility of foundation funding for her project:

> My ministry is anxious to get about two million low cost radio receivers so that our rural programmes and programmes on family planning could reach all parts of the country. I have written to FAO and UNESCO to find out whether we could get any assistance from them in getting cheap radio sets which our farmers could afford.

In the same letter she further outlined a program begun at her behest thus,

> With the help of the Government of West Germany, we have expanded our experimental Delhi Television Centre. Apart from educational

programmes during school hours, we have an hour's daily show of entertainment and social education programmes. ALL the TV sets at present are either in schools or community tele-clubs and a few in hospitals. Some are being put up for sale to the public. All this is in New Delhi and Old Delhi. However, we find that the TV range also extends to six hundred villages, out of which over three hundred have electricity. I feel that this is a wonderful opportunity to try out programmes of better farming methods as well as family planning through TV, since the visual impact is so much greater than the spoken word. We are trying to persuade village panchayats to buy TV sets. Due to the extreme shortage of foreign exchange, it is not possible to import more at the present moment. It would naturally expedite mattes if we could have help on this from any organization interested in family planning or in increased agricultural production.[72]

UNESCO showed little interest in her radio project. But she succeeded in interesting the UN in something ever more exciting – television. But India and Indira needed more than a toy television transmitter and the goodwill of UNESCO once she became the prime minister herself. Early in 1968, inaugurating a sounding rocket program in Trivandrum, Mrs. Gandhi confidently declared, "We envisage Television as an aid to education and development and as a force for national integration." May all the gods help![73]

Poor Indians, can't let them have nukes, and we can't get the clouds to cooperate. They have so little money, so little left to buy our fertilizer and insecticide factories. Maybe they can watch television! Works well back home, doesn't it? The Connecticut Yankee needed to do something, for he feared Merlin. "Could we help her to launch a major substitute?" Ambassador Bowles wondered.[74]

And no sooner had he thought so, the earth roared, the demi-gods applauded, the great gods clapped their hands, and divine perfumes, powders, and flowers rained from the heavens. Thus it was that the Connecticut Yankee in Indira's court helped America shape the Indian space program, the INSAT series of satellites in particular.

To complete my own story of how I came to write this account, after completing the required courses in the history and sociology of science, I did field research to get away from wintry cold Philadelphia. I found my way to tropical Kerala, researching the Aranmula mirror. There I met a woman, one of the Cochin Jews, and she told me an interesting version of the rabbi from Cracow story.[75]

Rabbi Eisik had a dream; the dream enjoined him to proceed to Prague where he would discover a hidden treasure buried under the bridge. But the rabbi, being poor, could not leave for Prague. He continued his prayers with additional fervor. The second night, he had yet another dream, clearly a continuation of the previous episode.

In this dream, he journeyed to Prague. Digging under the bridge he found only trash but no treasure. He was arrested by the captain of the guards and starved to death in prison. His desire for food at the moment of death caused him to be reborn to a rich Jewish merchant in Prague. He (one may, indeed, ask who) saw all this in his dream.

The merchant, disturbed by growing signs of anti-Semitism, emigrated to America with the reborn Rabbi, now but a child of seven years with the most wonderful dark curly hair, a winning smile, and a pair of twinkling eyes. The merchant's family passed through Ellis Island, had their names excised and anglicized. Our rabbi, called Isaac in the new world, grew up on the Lower East Side of Manhattan and eventually took over his father's small business. He, Isaac, naturally, enjoyed the fruits of his actions as the wise and holy rabbi of Cracow. His business prospered beyond all expectations. The American dream came true. In a few years, he dominated the entire garment fasteners industry and came to be Isaac, the zipper king. Brimming with wealth, he, the former Rabbi Eisik, decided to endow a chair in religious studies at Columbia University. At the ceremonial inaugural, he heard a professor of religion from the University of Chicago talk about Martin Buber, Heinrich Zimmer, and their story of Rabbi Eisik from Cracow. This reference had the rabbi wide awake.

He pinched himself. Yes, he was still a poor rabbi in Cracow. On the third morning of the strange dreams, Rabbi Eisik felt a vague sense of guilt. The American dream bothered him. "Why wasn't I reborn in the Holy Land?" thought he. So, he prayed extra hard on the third day. In a weak voice, trembling with hunger, the rabbi spoke of the impending rise of demonic forces in Europe and urged his congregation to move to the New World. Then, eager to see what he would dream on the third night, the Rabbi had an early supper and retired to his cold bedroom.

The third night, not surprisingly, Rabbi Eisik of Cracow dreamt yet another episode. It resumed with Isaac, the zipper king, endowing a religious studies chair at Columbia. And the professor from University of Chicago concluded her tale of the rabbi from Cracow with the rabbi obtaining the pot of gold in the hearth. Then, she alluded to an amusing version of the story authored by Woody Allen, in which the rabbi once again sets off to Prague to find the treasure:

> Two years later, he (the rabbi) was found wandering the Urals and emotionally involved with a panda. Cold and starving, the Rev was taken back to his house, where he was revived. . . . After telling the story, the Rabbi rose and went into his bedroom to sleep. . . . Three days later, he was back wandering in the Urals again, this time in a rabbit suit. . . . The above small masterpiece amply illustrates the absurdity of mysticism.[76]

Isaac, the zipper king, was furious, and walked up to Woody Allen, who was a member of the audience, and accused him of slander and shameless parody. Woody Allen rolled his eyes, shrugged his shoulders, raised his hands defensively, and spurted out, "Hey! Relax, I gotta make a living! I know! I know! I shouldn't have said mysticism was absurd. But, then again, this is your dream, not mine. If you have problems, talk to my attorney." So, Rabbi Eisik called Alan Dershowitz at Harvard, who convinced him to sue the professor from Chicago, the Universities of Chicago and Columbia, and various funding agencies of mythical research for violation of cultural and intellectual property rights. The rabbi litigated. And he won an unprecedented out-of-court settlement worth several million dollars from the academies and the mystical funding agencies. He also obtained a solemn promise from the professor not to steal any more Jewish stories. "Go get those Hindoos. They got oceans full of stories to spare!" advised the rabbi, in parting.

On the fourth morning the rabbi woke up, still in the ghetto in Cracow. Inspired by the dream professor's story, he dug in the corner of his hearth. Underneath years of accumulated trash, he found the treasure, a pot of gold. With his newfound wealth, he bought the most expensive house in Cracow. Being a reflective man, he told his congregation, "Listen! the treasure you seek is buried either in the heart or the hearth." And he added after a brief pause, "home is where the hearth with the treasure is." He lived happily (what is happiness?) ever after.

My story-sharing informant was seeking happiness. She was Indian of the Jewish faith and had been wandering through India looking for her guru. She said she heard this curiously Indian version of the rabbi from Cracow story from a Polish journalist in Tel Aviv. As it so happened, the journalist had spent considerable time in the 1930s and 1940s with the great Indian sage Ramana Maharishi in Arunachala. She told the story in a strange combination of Malayalam, Hebrew, English, and French. One can almost sense the spices of south India in the Hassidic tale heard on the Malabar coast, under the shade of cashew nut trees, fanned by the cool breeze of gently swaying coconut fronds carrying the aroma of tender pepper.

Her family had emigrated to their *punya bhoomi*, holy land, Israel, in 1952 only to find themselves homesick for the motherland, *mathru bhoomi*, India. So they returned, a year later, to Kerala. The fact that Nehru's India refused to entertain diplomatic relations with Israel had made it only more difficult to stay in Israel. A few years later, after securing a reliable supply of spices, saris, and Indian films through an Armenian trader in Madras, her family emigrated to Israel once again. Is there ever an easy pilgrimage?

And thus, I myself returned to Bharatvarsha after my formative sojourn in Philadelphia researching how art, history, philosophy, religion, culture – and of course transcontinental politics – shaped the extraordinary career of science in independent India.

Notes

1 In general, historical scholarship on and in India remains sequestered, as it were, in the colonial cantonment, secure in its Anglo-Indian mores. As a consequence, sources for the writing of Indian history after independence remain unexplored. In this dissertation, I have used North American archives to gain access to sources in India. One may note that the Nehru dynasty has controlled, rather tightly, the production of history in independent India. For a sample of the rich density of documentation available to an Indian historian, see the National Security Files (country:India) at the LBJ Library, Austin, Texas.

   More recently David Arnold has followed a long line of historians speculating on the causes and effects of famine in colonial India. Walter Rostow in private communication to Raja Rao has confirmed that LBJ looked forward to weekly briefings on the Indian monsoon and that he preferred the maps to tables. Raja Rao Archives, Austin, Texas. See also NSC.94/1: India's fall to Communism "would mean that for all practical purposes all of Asia would have been lost, . . . a serious threat to the security position of the United States."

2 Since the remote Indo-European past, Indians have prayed to the gods for the rain. Perhaps the oldest such prayer, still commonly chanted every day before a meal, is from the Atharvan Veda, (iv.15.6 and 9) "*yanty nadayo varshantu parjanyah*" "O! Parjanya, . . . let abundant rain come!" See also the later Chandogya Upanishad 2.2.2: "Let one meditate on the five fold Lord the Harmonious, in the rain, Pradyumna in the wind, that brings the rain clouds, Vasudeva in the gathered clouds, Narayana in the raining, Anirudhha in the thunder and lightning" (*Sacred Books of the East*).

3 It was not uncommon practice amongst post-colonial Anglophone Indian women to requisition American undergarments. For instance, Indira Gandhi had Dorothy Norman send her American bras through Pupul Jayakar. Dorothy Norman, a New York socialite, was also sympathetic to Gandhi's requests for plastic surgery. See Dorothy Norman, *Indira Gandhi, Letters to Friend, 1950–1984* (San Diego: Harcourt Brace Jovanovich, 1985).

   In the late 1950s, Indian scientists pioneered research in inducing rain and had developed a line of research not seen favorably by American meteorologists. The Indians preferred "cool" seeding of clouds in contrast to the Americans, who wanted to play with "hot" seeding. A history of Indian meteorology waits to be written.

   The Sanskrit quotation is from the Gita. Translations are mine, unless otherwise stated, and are selected to minimize the trauma of immigration into another language. See Ekendranath Ghosh, *Studies on Rigvedic Deities: Astronomical and Meteorological* (New Delhi: Cosmo Publications, 1983), for a non-orientalist perspective on the Rg Veda.

   Varuna, the archetypal Indo-European king, is not only the keeper of the waters but also the guardian of Truth. Thus, throughout Indian history there is a persistent relationship between royalty, virtue, and rainfall. See for instance in the *Manimekalai*, a Tamil Buddhist legend from circa 200 AD, "if the king swerves ever so little from righteousness, . . . rainfall will diminish."

4 The Sanskrit quotation is from the Taittriya Upanishad. See the oeuvre of Ananda K. Coomaraswamy, especially his *Spiritual Authority and Temporal Power in the Indian Theory of Government* (New Delhi: Munshiram Manoharlal, 1978), for a detailed exposition of Indian ideas of hospitality, power, etc. An exhaustive bibliography is available in Roger Lipsey, *Coomaraswamy: His Life and Work, Vols. 1–3* (Princeton, NJ: Princeton University Press, 1977). The Rg Veda has a number of hymns on making rain. See 7.10, 10.98 for instance.

5 A detailed history of weather modification is yet to be written. For an excellent essay on the subject, see James Fleming's essay, "Cloud Wars: Weather Modification and the US Military, 1947–1977." I am much indebted to Oswald Spengler's seminal work *The Decline of the West* (New York: Modern Library, 1965). The German quotations are from Goethe's *Faust*. Bayard Taylor's translation is still among the best. See Johann Goethe, *Faust*, revised and edited by Stuart Atkins, translated by Bayard Taylor (New York: Collier Books, 1962). I am thankful to T. P. Hughes for pointing out this translator.

6 The *vasordara* is frequently depicted in Indian sculptures and is thus a part of the Indic visual imagination. The gift is verily a sacrifice and hence the utterance "This is but smoke" accompanying gift-giving. Once again A. K. Coomaraswamy's expositions remain unsurpassed even after several decades. See Roger Lipsey volume.

7 In the Patanjali Yoga sutras, a certain type of yogic rapture is referred to as the peace of the rain cloud of dharma. The verse quoted is Buddhist in origin, from Raja Rao's second volume in *The Chessmaster* trilogy. W. P. Lehman has compared Raja Rao's *The Chessmaster* to Goethe's *Faust*.

8 The Taitriya Samhita (8.4, 2.,1, for instance) declares that "In the beginning, the gods were mortal; they became divine and immortal through sacrifice; they live by gifts from the earth, as men live by gifts from heaven." This exchange of gifts between humans and gods was, as it were, the key to prosperity according to ancient Indian thought. Fritz Staal at University of California, Berkeley, has done some of the most provocative work on the nature of Vedic ritual. See his *Agni – The Vedic Ritual of the Fire Altar* (Berkeley: Asian Humanities Press, 1983) and the controversy stirred by it for the meaninglessness of ritual scholarship.

9 Hughes suggests that "In asking the moment to linger, Faust was asking that the joy of creation be prolonged. . . . To ask the creative moment to linger was not only to lose the wager but to demand the impossible." See T.P. Hughes, "Elmer Sperry and Adrian Leverkuhn: A Comparison of Creative Styles," in *Springs of Scientific Creativity: Essays on Founders of Modern Science*, edited by Rutherford Aris, et al. (Minneapolis: University of Minnesota Press, 1983), p. 200. The fourth brahmin in the Vedic sacrifice sits silently. T. J. Ferguson of the Institute of the American West is engaged in pioneering anthropological work in the American southwest. His remarkable unpublished essay on the ongoing repatriation of Zuni war gods provides a brief account of a post-traditional anthropology. The German quotation is once again from Goethe's *Faust*. The Sanskrit quotations are frequently encountered in multiple texts. Fritz Staal, *Exploring Mysticism: A Methodological Essay* (Berkeley: University of California Press, 1975). On sacrifice: see Sathapatha Brahmana 8.6., 1, 10.

10 Rachel Carson, *Silent Spring* (Boston: Houghton Mifflin, 1962). See also Patricia Hynes, *The Recurring Silent Spring* (New York: Pergamon Press, 1989). Douglas Ensminger of Ford Foundation was in the forefront persuading India's first president, Rajendra Prasad, of the non-toxicity of fertilizers and insecticides. Ananda Coomaraswamy, *Yaksas: Essays in the Water Cosmology* (New Delhi: Oxford University Press and IGNCA, 1993), pp. 113–127. He argues that "the ideal of kingship embodied in the original conception of Varuna may be said to have persisted in Indian culture up to the present day" (115). If not anything else, the royal umbrella survives to this day.

11 A thick history of the 1962 conflict with China is not yet available. My observations are based on extensive interviews and reading of newspapers and archives. See also N. Maxwell, *India's China War* (London: Cape, 1970), for a journalistic account. See Jawaharlal Nehru, *Freedom's Daughter: Letters Between Indira*

Gandhi and Jawaharlal Nehru, edited by Sonia Gandhi (London: Hodder and Stoughton, 1989). Also see F. L. Jawhar, *The Tezkereh al Vaki:at, or Private Memoirs of the Moghul Emperor Humayun*, translated by Charles Stewart (New Delhi: Kumar Bros., 1970).

12 Dary entry, 20 February 1965, Chester Bowles Papers, Yale University, p. 86. Also see J. P. Dalvi, *Himalayan Blunder: The Curtain-Raiser to the Sino-Indian War of 1962* (Mumbai: Thacker, 1969).

13 See Howard B. Schaffer, *Chester Bowles: New Dealer in the Cold War* (Cambridge: Harvard University Press, 1993).

14 The Bowles children were particularly tired because their parents insisted on rushing them through Parisian museums, giving them, as it were, a stiff dose of Western civilization before going to India. Oral History, Chester Bowles Papers, Yale University, 1965, pp. 500–505. *Darpanantaha puram yatha . . .* Viveka Chudamani of Adi Sankara: "just as there is a reflection of the city in a mirror, the universe is reflected in the Self."

15 Oral History, Chester Bowles Papers, Yale University, 1965, pp. 500–505. Bowles remembered that President Truman was appalled at the thought of anyone wanting to go there. According to Bowles, Truman said, "Well, I thought India was pretty jammed with poor people and cows wandering around the streets, witch doctors, and people sitting on hot coals and bathing in the Ganges, and so on, but I did not realize that anybody thought it important." Also see Raja Rao, *The Cat and Shakespeare*, for an amusing exposition of the mother cat in Indian philosophy.

16 Oral History, Chester Bowles Papers, Yale University, 1965, pp. 500–505. Also see Raja Rao, *The Serpent and the Rope* (Woodstock: Overlook Press, 1986).

17 Meister Eckhart, *Meister Eckhart*, edited by Franz Pfeiffer (Leipzig, 1857), Translated by C. de. B. Evans (London, 1924–1931), vol. 1, p. 85. Oral History, Chester Bowles Papers, Yale University, 1965, pp. 500–508. See the prolific work of H. W. Brands on the cold war, especially his *Inside the Cold War: Loy Henderson and the Rise of the American Empire, 1918–1961* (New York: Oxford University Press, 1991). Also see Dean Rusk, *As I Saw It* (New York: Norton, 1990) for background on the Truman-Nehru meeting.

18 The *dig-gajas*, the cardinal elephants, have been associated with clouds since Vedic times and perhaps even earlier as elements of a water cosmology. In Sanskrit poetics, elephants are often favorite images for clouds. Water sporting elephants, *jalebha*, are also seen in the Indian plastic arts through history. See P. N. Kane, *History of Sanskrit Poetics*, 4th edition (New Delhi: Motilal Banarasidass, 1971). The celestial elephants are also seen in paintings and sculptures throughout the Indic world.

19 I attempt to write a Bhaktinian history, dialogic and heteroglossic in nature. See Mikhail Bhaktin, *The Dialogic Imagination*, edited by Michael Holquist (Austin: University of Texas Press, 1981).

20 C. Subramanium, *Matsya Purana* (New Delhi: Nag, 1983), Sacred Books of the Hindus, vol. 17.

21 On the famine of 1943 in Bengal, see Paul Greenough, *Prosperity and Misery in Modern Bengal: The Famine of 1943–1944* (New York: Oxford University Press, 1982). A detailed history of post-independence India free of Nehru hagiography is yet to be written. This is a modest beginning in that direction.

22 The quotation is a famous one of Homi Bhabha. See his *Science and the Problems of Development* (Mumbai: Atomic Energy Establishment, 1966). Also see Chapter 3 of *Goods and Gods* for detailed study of Bhabha. See Morarji Desai, *The Story of My Life*, (New Delhi: Macmillan, 1974–79), for an account of the

negotiations over the wartime loans to Britain. A history of the role of the Office of Munitions Control in the conduct of cold war waits to be written.

23 See US Atomic Energy Commission. Hewlett, Richard, *Atoms for Peace and War, 1953–1961: Eisenhower and the Atomic Energy Commission* (Berkeley: University of California Press, 1989). Daniel Kevles, *The Physicists: The History of a Scientific Community in Modern America* (New York: Vintage Books, 1979).

24 Robert Anderson, *Building Scientific Institutions in India: Saha and Bhabha* (Montreal: Centre for Developing-Area Studies, McGill University, 1975).

25 The Gazette of India, 1 February 1966, no. 29 part 1 section 1, p. 71.

26 Vikram Sarabhai at Trombay, 25 January 1966, quoted in *Nuclear India*, February 1966, p. 7. As the immediate successor to Bhabha, Sarabhai delivered a touching ritual speech condoling the death of Homi Bhabha to all employees of the Indian Atomic Energy Commission who assembled in the open grounds in front of the Trombay nuclear reactor. The only scholarly work on Homi Jehangir Bhabha is a comparative study of Homi Bhabha and Meghnad Saha. See Robert S. Anderson, *Building Scientific Institutions in India: Saha and Bhabha* (Montreal: Centre for Developing-Area Studies, McGill University, 1975).

27 Homi Bhabha's grandfather had been a successful minister of education in the princely state of Mysore. R. Kulkarni, *Homi Bhabha: Father of Nuclear Science in India* (Mumbai: Popular Prakashan, 1969); P. R. Pisharoty, *C. V. Raman* (New Delhi: Publications Division, Ministry of Information and Broadcasting, Govt. of India, 1982).

28 G. Venkatraman, *Journey into Light: Life and Science of C.V. Raman* (Bangalore: Indian Academy of Sciences, 1988), p. 19; M. G. K. Menon, Biographical memoir citation XXX; See Suma Chitnis, *The Indian Academic Profession: Crisis and Change in the Teaching Community* (New Delhi: Macmillan, 1979); Homi Bhabha on TIFR in a letter to Droab Tata appealing for funds. Bhabha papers. Bombay.

29 George Greenstein, "A Gentleman of the Old School: Homi Bhabha and the Development of Science in India," *American Scholar*. (Summer 1992): 409–419.

30 See Shiv Visvanathan, *Organising for Science: The Making of an Industrial Research Laboratory* (New Delhi: Oxford University Press, 1985); G.Venkatraman, *Journey into Light: Life and Science of C.V. Raman* (Bangalore: Indian Academy of Sciences, 1988).

31 Homi Bhabha, *Nuclear India*, 6 February 1963, p. 6.

32 I have drawn extensively on the work of John Lewis and Xue Litai, *China Builds the Bomb* (Stanford: Stanford University Press, 1988), for this section.

33 NSF files, (country: India) LBJ Library, Austin, Texas.

34 The Chinese conquest of Tibet meant, among other things, that Mount Kailas, the abode of Kubera, the daimon of wealth in the Indian tradition, went into Chinese hands. Soon enough, India began losing the nine treasures of Kubera. Liu, Xinru, *Early Commercial and Cultural Exchanges Between India and China . . .* PhD Dissertation (Philadelphia: University of Pennsylvania, 1985). NSF Files (country: India) LBJ Library, Austin, Texas.

35 NSF files, (country: India) LBJ Library, Austin, Texas. See also C.S. Jha, *From Bandung to Tashkent: Glimpses of India's Foreign Policy* (Chennai: Sangam Books, 1983).

36 NSF files, (country: India) LBJ Library, Austin, Texas.

37 Oral history, Chester Bowles Papers, Yale University.

38 Raja Rao, *The Chessmaster and His Moves* (New Delhi: Vision Books, 1988), p. 206. The Chinese war was personally humiliating to Nehru. It decreased his stature as a statesman and eclipsed much of his vision. The Chinese nuclear

program makes an interesting comparison with the Indian space program. John Lewis and Xue Litai, *China Builds the Bomb* (Stanford: Stanford University Press, 1988). Confidential and personal letter from Bowles to Senator Robert. F. Kennedy, 19 October 1965. Chester Bowles Papers, Yale University.

39 It is apparent that the younger generation of leadership that took over after 1966, Indira Gandhi in politics and Vikram Sarabhai in technology, slowed the Indian nuclear weapons program on account of American persuasion and opted to expand the Indian space program and green revolution. Diary entry, 10 September 1964. Chester Bowles Papers, Yale University. The phrase cloud war is from James Fleming's unpublished essay.

40 On how India copes with droughts, see Martha Alter Chen, *A Gamble on the Monsoon: Coping with Seasonality and Drought in Western India*, PhD Dissertation (Philadelphia: University of Pennsylvania, 1989). The quotations are from Chester Bowles' diary, Chester Bowles Papers, Yale University.

41 Ibid. Only the bodies of fifteen rhesus monkeys were found at the crash site. The film *Mountain* dramatizes a 1950 Air India plane crash on Mt. Blanc.

42 Based on extensive readings of Indian newspapers. See also diary entries Jan-Feb, 1966, Chester Bowles Papers, Yale University. For India-Pakistan wars, see Sumit Ganguly, *The Origins of War in South Asia: Indo-Pakistani Conflicts Since 1947* (Boulder: Westview Press, 1986).

43 This is a Rg Vedic hymn of creation. The creator is ritually sacrificed by the created beings. This archaic behavior seems to replay itself endlessly in India.

44 Based on extensive material in the Rockefeller Archives. See also diary entry, 23 January 1966, Chester Bowles Papers, Yale University.

45 I am deeply indebted to James Capshew for directing me to Jim Fleming, who in turn, has been extremely generous in sharing his expertise on the history of weather modification. Diary entry, 19 July 1964. Chester Bowles Papers, Yale University. Information on Project GROMET and POPEYE is to be found, among other places, in National Security Files, (India), LBJ library, Austin, Texas. NSF country files, India, Boxes 131, 132, Minutes and Memos, Volume ix.

46 For a general introduction to weather modification, see Committee on Atmospheric Sciences, National Research Council, *Weather & Climate Modification: Problems and Progress*Washington, D.C.: National Academy of Sciences, 1973).

47 Based on materials in LBJ library and Rockefeller Archives. See diary entries, 6 February, 25 February, and 6 April 1966, Chester Bowles Papers, Yale University. C. Subramaniam was the chief negotiator on the Indian side for much of the green revolution. See his *India of My Dreams* (New Delhi: Orient Longman, 1972) and *The New Strategy in Indian Agriculture: The First Decade and After* (New Delhi: Vikas, 1979). See also Francine Frankel, *India's Green Revolution, Economic Gains and Political Costs* (Princeton, NJ: Princeton University Press, 1971).

48 Operation GROMET files. See NSF country files, India, Boxes 131, 132, Minutes and Memos, Volume ix. LBJ Library, Austin, Texas. Apparently LBJ enjoyed speeding down a steep hill with his guests in a specially built convertible on his ranch. At the bottom of the hill was a pond. As the car reached top speed, he would yell at his guests that the brakes did not work. It would turn out that the car was amphibian. LBJ just loved watching how his guests reacted. Many, including Dean Rusk, I am told, bailed out, much to their embarrassment.

49 See Senate Committee on Foreign Relations, Subcommittee on Oceans and International Environment, *Prohibiting Military Weather Modification: Hearings on S.R. 281, 92nd Cong., 2nd Sess.*, 1972. Senate Committee on Foreign Relations, Subcommittee on Oceans and International Environment, *Weather Modification: Hearings, 93rd Cong., 2nd Sess.*, 1974. Senate Committee on Foreign

Relations, Subcommittee on Oceans and International Environment, *Prohibition of Weather Modification as a Weapon of War: Hearings on H.R. 28, 94th Cong., 1st Sess.*, 1975. Also see NSF country files, India, Boxes 131, 132, Minutes and Memos, Vol. ix.

50 NSF country files, India, Boxes 131, 132, Minutes and Memos, Vol. ix.

51 The Sanskrit quotation is from Rg Veda: *Accha vadam tavasam.*

52 My informants were appalled at the utter disregard for the scientific method shown by the US military in Operation GROMET. A detailed history of this scientific controversy is awaited.

53 On development talk, see for example, Oral History, pp. 515–526, Chester Bowles Papers, Yale University. Stalin was rather popular among certain political circles in India. For instance, the chief of the Dravida movement, M. Karunanidhi, named his eldest son and political heir after Stalin.

54 Vikram Sarabhai, "there is no hope for those who wish to advance step by step following outmoded methods." 2 February 1968 in *Nuclear India*, vol. 6, no. 6–7, February–March, 68.

55 A classic example is the encounter between the president of Westinghouse and the Indian prime minister early in 1949. The State Department had to severely reprimand the entrepreneur. Several distinguished members of the early Indian Foreign Service were drawn from the Malayalee nobility, who had grown up with stories of the Zamorin and Vasco da Gama.

56 John Foster Dulles, in a talk with US Embassy staff in New Delhi, April, 1953. Chester Bowles diary entry, 10 September 1963. Chester Bowles Papers, Yale University.

57 Oral History, p. 520, Chester Bowles Papers, Yale University.

58 The brimming pot, the *poorna-kalasham*, is the symbol of auspiciousness in Indic culture, occurring not only in the subcontinent but also Southeast Asia. Often a lotus blooms out of such a pot. To this day, a guest (even Chester Bowles was honoured thus) is welcomed with an offering of the brimming vase, the *poorna-kalasham*. Kubera, the chief daimon of wealth, lives in the northern quarter.

59 Raja Rao, *The Chessmaster and His Moves* (New Delhi: Vision Books, 1988).

60 There is a universal view in the Indic tradition that the pupil has total intellectual freedom to question the teacher. The greatest texts of Indian philosophy are structured as a series of questions and answers. In fact, it is only by asking the right question that the right answer will be provoked. And the right answer would have the efficacy of an "Act of Truth." Ananda Coomaraswamy suggests that "There may be a parallel here to the primary importance attached in the Grail Quest to the asking of the right questions; in India, magical efficacy is attributed to a statement of the truth." Ananda Coomaraswamy, *Yaksas: Essays in the Water Cosmology* (New Delhi: Oxford University Press and IGNCA, 1993), p. 127.

61 I use such sections to convey a taste of a different kind of historiography that I grew up with. Bindumati, the courtesan, by the mere assertion of the truth, showed the great emperor Ashoka that water in the mighty River Ganga flowed upstream.

62 Andre Malraux, *Anti-Memoirs*, translated by Terence Kilmartin (New York: Holt, Reinhart, and Winston, 1968). Homi Bhabha quoting Gandhi in a broadcast on AIR, 1 June 1964.

63 Gandhiji wrote, "The idea of swadeshi is of great importance and the progress of the country in dharma is bound up with it. . . . Religious leaders, too, should follow the rule of swadeshi. They have plenty of time on hand. They should take to the spinning-wheel and spin and thus set an example to their followers." 6

July 1919. A biography of Maurice Frydman waits to be written. A brief sketch of his life may be found later in the manuscript. M. K. Gandhi, *Hind Swaraj or Indian Home Rule* (Ahmedabad: Navjivan, 1984, reprint of 1938 edition).

64 Andre Malraux, *Anti-Memoirs*, translated by Terence Kilmartinv (New York: Holt, Rinehart & Winston, 1968). For the Afro-Asian conference, see *Romulo, Carlos Pena, Meaning of Bandung* (Chapel Hill: University of North Carolina Press, 1956). George McTurnan Kahin, *The Asian-African Conference, Bandung, Indonesia, April 1955* (Ithaca: Cornell University Press, 1956). Richard Wright, *The Color Curtain: A Report on the Bandung Conference* (Cleveland: World Pub. Co., 1956).

65 M. K. Gandhi, *Hind Swaraj or Indian Home Rule* (Ahmedabad: Navjivan, 1984, reprint of 1938 edition). The philosopher Wittgenstein invented a sewing machine.

66 Homi Bhabha, *Science and the Problems of Development* (Mumbai: Atomic Energy Establishment, 1966). A. K. Coomaraswamy, *Art and Swadeshi* (Chennai: Ganesh, 1911), pp. 3–4. S. Gopal, (ed.), *Jawaharlal Nehru: An Anthology* (New Delhi: Oxford University Press, 1980).

67 Speechifying in Indian English lends a particular weight to the action of speaking otherwise unavailable in English. See Rajmohan Gandhi's excellent biography, *Patel: A Life* (Ahmedabad: Navajivan Publishing House, 1990), for a sustained exploration of the tensions between Nehru and Patel. Rajmohan Gandhi's biography of Rajaji is also highly commended. Rajmohan Gandhi, *The Rajaji Story* (Chennai: Bharathan, 1978–1984). It is probably the best biography of any twentieth-century Indian figure.

68 Benjamin Lee Whorf, *Language, Thought, and Reality; Selected Essays*, edited by John B. Carroll (Cambridge: Technology Press of MIT, 1956). See Ananda K. Coomaraswamy, *Spiritual Authority and Temporal Power in the Indian Theory of Government* (New Delhi: Munshiram Manoharlal, 1978), cited in Lipsey.

69 It may seem strange to a Western mind that more often than not activism in India takes the form of speech. A primordial belief in the magical potency of the word continues to pervade India even today. To utter the Word, the mantra is to create. "He said *bhuh*, and the earth was. He created the waters by means of the word." Sathapatha Brahmana 11.1.6.3 and 6.1.1.9, respectively. The early history of the Congress in India was the history of formulating verbose petitions. Homi Bhabha marshaled the powers of the words "Atoms for Peace" effectively in the 1950s. His success at the three Geneva conferences inspired Vikram Sarabhai to seek the chair of the 1968 Vienna conference on the peaceful uses of outer space. "Peaceful uses of outer space" had all the right connotations of a Sanskrit mantra. The ritual incantation of peace had become a feature of Indian foreign policy by mid-1960s. "He (Nehru) was so important to India and India's survival, so important to all of us, that if he did not exist – as Voltaire said of God – he would have to be invented." Dean Acheson, *Present at the Creation* (New York: Norton, 1969), p. 62. Office Files of Asst. (South Asia, 1945–1950) Secretary of State, National Archives, Washington, D.C. See, for example, Olaf Caroe, *The Pathans, 550 BC–AD 1957* (Karachi: Oxford University Press, 1983). See also Dost Muhammad Khan Kamil, *On a Foreign Approach to Khushhal, a Critique of Caroe and Howell* (Peshawar: Maktabah-i-Shaheen, 1968) for information on Caroe. *Vagvi brhati . . . vagvi brhman.* Brahadaranyaka Upanishad. 1.3.19–20.

70 "The word is not only a big deal but verily gold" declares the Chandogya Upanishad. An extraordinarily imaginative individual, Bowles had created an innovative advertising agency, Benton & Bowles, at the height of the Depression by

specializing in the new field of radio advertising. He was one of the key figures in the "radio days" of the 1930s, launching such seminal programs as the Maxwell House Showboat. It is said that his advertising was so persuasive that people lined up at the docks in Memphis to see the showboat come and were sorry to hear it was all make-believe. First lured into marketing by a weekly offer of $25, Bowles was making a quarter-million dollars a year by the mid-1930s. His biographer notes, "the success of the agency was particularly remarkable because it came in the worst of the Depression." On speech-making, see, for example, Letter from Chester Bowles to Dean Rusk, 30 September 1963, Chester Bowles Papers, Yale University. For his background in advertising, also see Howard B. Schaffer, *Chester Bowles: New Dealer in the Cold War* (Cambridge: Harvard University Press, 1993), p. 12. On *The American Reporter*, based on interviews with readers of *The American Reporter* in Kammangudi, May, 1989.

71 Letter from Chester Bowles to Indira Gandhi, 21 May 1965. Chester Bowles Papers, Yale University Archives. Vacaspati, literally the master of words, is another name for the divine artisan, Viswakarman. In the Rg Veda, perhaps the oldest religious hymn known to man, Viswakarma, the master craftsman, is addressed as Vacaspati. The poet is also an engineer.

72 Indira Gandhi to Dorothy Norman, 5 November 1965 in Dorothy Norman, *Indira Gandhi: Letters to an American Friend, 1950–1984* (San Diego: Harcourt Brace Jovanovich), p. 113.

73 Ibid. American experts like Wilbur Schramm also met with Indira Gandhi and convinced her of the need to invest in broadcast communication technology.

74 In the end, American scientists have indeed succeeded in cloud-seeding technology. Today at least fifteen states approve of cloud-seeding, while chanting and dancing for rain is practiced in two states. In the meanwhile, in India, we mostly resort to prayers for rains. See memo by Bob Brooks to Chester Bowles, undated but circa 1967, in Chester Bowles Papers. Also see diary, p. 155, Chester Bowles Papers, Yale University.

75 Interviews with Rachel Abraham, Aranmula, February 1986.

76 Woody Allen, *Getting Even* (New York: Random House, 1972), pp. 52–56.

Works cited

Acheson, Dean. *Present at the Creation*. New York: Norton, 1969.

Allen, Woody. *Getting Even*. New York: Random House, 1972.

Anderson, Robert. *Building Scientific Institutions in India: Saha and Bhabha*. Montreal: Centre for Developing-Area Studies, McGill University, 1975.

Bhabha, Homi. *Science and the Problems of Development*. Mumbai: Atomic Energy Establishment, 1966.

Bhaktin, Mikhail. *The Dialogic Imagination*. Ed. Michael Holquist. Austin: University of Texas Press, 1981.

Brands, H.W. *Inside the Cold War: Loy Henderson and the Rise of the American Empire, 1918–1961*. New York: Oxford University Press, 1991.

Caroe, Olaf. *The Pathans, 550 B.C.-A.D. 1957*. Karachi: Oxford University Press, 1983.

Carson, Rachel. *The Silent Spring*. Boston: Houghton Mifflin, 1962.

Chen, Martha Alter. *A Gamble on the Monsoon: Coping with Seasonality and Drought in Western India*. PhD Dissertation. Philadelphia: University of Pennsylvania, 1989.

Coomaraswamy, Ananda K. *Art and Swadeshi*. Chennai: Ganesh, 1911.

————. *Spiritual Authority and Temporal Power in the Indian Theory of Government*. New Delhi: Munshiram Manoharlal, 1978.

————. *Yaksas: Essays in the Water Cosmology*. New Delhi: Oxford University Press and IGNCA, 1993.

Dalvi, J.P. *Himalayan Blunder: The Curtain-Raiser to the Sino-Indian War of 1962*. Mumbai: Thacker & Co., 1969.

Desai, Moraji. *The Story of My Life*. New Delhi: Macmillan Reference, 1974–79.

Eckhart, Meister. *Meister Eckhart*. Ed. Franz Pfeiffer. Trans. C. de. B. Evans. Vol. 1. London: J.M. Watkins, 1924–31.

Frankel, Francine R. *India's Green Revolution, Economic Gains and Political Costs*. Princeton: Princeton University Press, 1971.

Gandhi, M.K. *Hind Swaraj or Indian Home Rule*. Ahmedabad: Navjivan, reprint, 1984.

Ganguly, Sumit. *The Origins of War in South Asia: Indo-Pakistani Conflicts Since 1947*. Boulder: Westview Press, 1986.

Ghosh, Ekendranath. *Studies on Rigvedic Deities: Astronomical and Meteorological*. New Delhi: Cosmo Publications, 1983.

Goethe, Johann. *Faust*. Ed. Stuart Atkins. Trans. Bayard Taylor. New York: Collier Books, 1962.

Greenough, Paul. *Prosperity and Misery in Modern Bengal: The Famine of 1943–1944*. New York: Oxford University Press, 1982.

Greenstein, George. "A Gentleman of the Old School: Homi Bhabha and the Development of Science in India." *American Scholar*. (Summer 1992): 409–419.

Hewlett, Richard. *Atoms for Peace and War, 1953–1961: Eisenhower and the Atomic Energy Commission*. Berkeley: University of California Press, 1989.

Hughes, T.P. "Elmer Sperry and Adrian Leverkuhn: A Comparison of Creative Styles." *Springs of Scientific Creativity: Essays on Founders of Modern Science*. Eds. Aris Rutherford et al. Minneapolis: University of Minnesota Press, 1983.

Hynes, Patricia. *The Recurring Silent Spring*. New York: Pergamon Press, 1989.

Jawhar, F.L. *Private Memoirs of the Moghul Emperor Humayun*. Trans. Charles Stewart. New Delhi: Kumar Bros., 1970.

Jha, C.S. *From Bandung to Tashkent: Glimpses of India's Foreign Policy*. Chennai: Sangam Books, 1983.

Kahin, George McTurnan. *The Asian-African Conference, Bandung, Indonesia, April 1955*. Ithaca: Cornell University Press, 1956.

Kane, O.N. *History of Sanskrit Poetics*. 4th ed. New Delhi: Motilal Banarasidass, 1971.

Kevles, Daniel. *The Physicists: The History of a Scientific Community in Modern America*. New York: Vintage Books, 1979.

Khan, Kamil Dost Muhammad. *On a Foreign Approach to Khushhal, a Critique of Caroe and Howell*. Peshawar: Maktabah-i-Shaheen, 1968.

Kulkarni, R.P. *Homi Bhabha: Father of Nuclear Science in India*. Mumbai: Popular Prakashan, 1969.

Lewis, John, and Xue Litai. *China Builds the Bomb*. Stanford: Stanford University Press, 1988.

Lipsey, Roger. *Coomaraswamy: His Life and Work, Vols. 1–3*. Princeton, NJ: Princeton University Press, 1977.

Malraux, Andre. *Anti-Memoirs*. Trans. Terence Kilmartin. New York: Holt, Reinhart, and Winston, 1968.

Maxwell, N. *India's China War*. London: Cape, 1970.

Nehru, Jawaharlal. *Freedom's Daughter: Letters Between Indira Gandhi and Jawaharlal Nehru*. Ed. Sonia Gandhi. London: Hodder and Stoughton, 1989.

Norman, Dorothy. *Indira Gandhi, Letters to Friend, 1950–1984*. San Diego: Harcourt Brace Jovanovich, 1985.

Pisharoty, P.R. *C.V. Raman*. New Delhi: Publications Division, Ministry of Information and Broadcasting, Govt. of India, 1982.

Rao, Raja. *The Serpent and the Rope*. Woodstock: Overlook Press, 1986.

———. *The Chessmaster and His Moves*. New Delhi: Vision Books, 1988.

———. *The Cat and Shakespeare*. New Delhi: Penguin Books India, 2014.

Romulo, Carlos Pena. *Meaning of Bandung*. Chapel Hill: University of North Carolina Press, 1956.

Rusk, Dean. *As I Saw It*. New York: Norton, 1990.

Schaffer, Howard B. *Chester Bowles: New Dealer in the Cold War*. Cambridge: Harvard University Press, 1993.

Spengler, Oswald. *The Decline of the West*. New York: Modern Library, 1965.

Staal, Fritz. *Exploring Mysticism: A Methodological Essay*. Berkeley: University of California Press, 1975.

———. *Agni-The Vedic Ritual of the Fire Altar*. Berkeley: Asian Humanities Press, 1983.

Subramaniam, C. *India of My Dreams*. New Delhi: Orient Longman, 1972.

———. *The New Strategy in Indian Agriculture: The First Decade and After*. New Delhi: Vikas, 1979.

———. *The Matsya Purana*. Vol. 17, 2 Vols. Sacred Books of the Hindus. New Delhi: Nag Publishers, 1983.

Venkatraman, G. *Journey into Light: Life and Science of C.V. Raman*. Bangalore: Indian Academy of Sciences, 1988.

Visvanathan, Shiv. *Organising for Science: The Making of an Industrial Research Laboratory*. New Delhi: Oxford University Press, 1985.

Whorf, Benjamin Lee. *Language, Thought, and Reality; Selected Essays*. Ed. John B. Carroll. Cambridge: Technology Press of MIT, 1956.

Wright, Richard. *The Color Curtain: A Report on the Bandung Conference*. Cleveland: World Publishing Company, 1956.

INDEX